Contents

Computer Algebra Systems in Secondary School Mathematics Education

INTRODUCTION 1
   Editorial Panel

ACTIVITIES (LOCATED THROUGHOUT BOOK)
   Lin McMullin
      Niantic, Connecticut

PART 1: PERSPECTIVES FOR ANALYZING CAS POTENTIAL 7
   Perspectives for Analyzing CAS Potential: Introduction 7
      Lin McMullin
         Niantic, Connecticut

   1   Algebra and Computer Algebra 9
       E. Paul Goldenberg
         EDC Center for Mathematics Education, Newton, Massachusetts

   ACTIVITY 1 31

   2   Theories for Thinking about the Use of CAS in
       Teaching and Learning Mathematics. 33
       M. Kathleen Heid
         Pennsylvania State University, University Park, Pennsylvania

   3   CAS as Pedagogical Tools for Teaching and Learning
       Mathematics 53
       Bernhard Kutzler
         Austrian Center for Didactics of Computer Algebra, Linz, Austria

   ACTIVITY 2 72

   4   Thinking out of the Box 73
       William G. McCallum
         University of Arizona, Tucson, Arizona

PART II: EXAMPLES OF CAS AT WORK IN THE CURRICULUM AND CLASSROOM 87

Examples of CAS at Work in the
Curriculum and Classroom: Introduction 87
Carolyn Kieran
University of Quebec at Montreal, Montreal, Quebec
Al Cuoco
EDC Center for Mathematics Education, Newton, Massachusetts

5   Promoting Pure Mathematics through Preliminary
Investigational Activities Using Computer Algebra 89
David Bowers
Suffolk College, Ipswich, United Kingdom

ACTIVITY 3 96

6   Classical Mathematics in the Age of CAS 97
Al Cuoco
EDC Center for Mathematics Education, Newton, Massachusetts
Ken Levasseur
University of Massachusetts at Lowell, Lowell, Massachusetts

7   Calculator-Based Computer Algebra Systems:
Tools for Meaningful Algebraic Understanding 117
Michael Todd Edwards
John Carroll University, University Heights, Ohio

ACTIVITY 4 135

8   Computing, Conjecturing, and Confirming
with a CAS Tool 137
Tim Garry
Copenhagen International School, Hellerup, Denmark

9   Technology Matters: An Invitation to
Generating Functions with CAS 151
Jeremy A. Kahan
University of Minnesota, Minneapolis, Minnesota
Terrence R. Wyberg
University of Minnesota, Minneapolis, Minnesota

Contents

ACTIVITY 5 .... .162

10 To CAS or Not to CAS? .... .163
    James E. Schultz
    Ohio University, Athens, Ohio

11 Task Design in a CAS Environment:
    Introducing (In)equations .... .173
    Nurit Zehavi
    The Weizmann Institute of Science, Rehovot, Israel
    Giora Mann
    The Weizmann Institute of Science, Rehovot, Israel

ACTIVITY 6 .... .192

PART III: EVIDENCE AND IMPLICATIONS FROM RESEARCH .... .195

Evidence and Implications from Research: Introduction .... .195
Rose Mary Zbiek
    Pennsylvania State University, University Park, Pennsylvania

12 Using Research to Influence Teaching and
    Learning with Computer Algebra Systems .... .197
    Rose Mary Zbiek
    Pennsylvania State University, University Park, Pennsylvania

ACTIVITY 7 .... .217

13 Initiating Students into Algebra with
    Symbol-Manipulating Calculators .... .219
    Tenoch Cedillo
    National Pedogogical University, Mexico City, Mexico
    Carolyn Kieran
    University of Quebec at Montreal, Montreal, Quebec

14 Algebra on Screen, on Paper, and in the Mind .... .241
    Paul Drijvers
    Freudenthal Institute, Utrecht, The Netherlands

ACTIVITY 8 268

15 Learning Techniques and Concepts Using CAS:
    A Practical and Theoretical Reflection 269
    Jean-Baptiste Lagrange
    IUFM de Reims, France, et Équipe DIDIREM, Université Paris, France

ACTIVITY 9 285

PART IV: CAS AND ASSESSMENT OF MATHEMATICAL UNDERSTANDING AND SKILL 287

CAS and Assessment of Mathematical
Understanding and Skill: Introduction 287
James T. Fey
University of Maryland, College Park, Maryland

16 What Should Students Record When Solving Problems with
    CAS? Reasons, Information, the Plan, and Some Answers 289
    Lynda Ball
    University of Melbourne, Victoria, Australia
    Kaye Stacey
    University of Melbourne, Victoria, Australia

ACTIVITY 10 304

17 Testing with Technology: Lessons Learned 305
    Raymond J. Cannon
    Baylor University, Waco, Texas
    Bernard L. Madison
    University of Arkansas, Fayetteville, Arkansas

18 Traditional Assessment and Computer Algebra Systems 329
    Lin McMullin
    Niantic, Connecticut

ACTIVITY 11 335

Computer Algebra Systems in Secondary School Mathematics Education

Computer Algebra Systems in Secondary School Mathematics Education

Edited by
James T. Fey, Chair
Al Cuoco
Carolyn Kieran
Lin McMullin
Rose Mary Zbiek

NATIONAL COUNCIL OF
TEACHERS OF MATHEMATICS

Copyright © 2003
THE NATIONAL COUNCIL OF
TEACHERS OF MATHEMATICS, INC.
1906 Association Drive, Reston, VA 20191-1502
(703) 620-9840; (800) 235-7566; www.nctm.org
All rights reserved
ISBN 0-87353-531-6

The publications of the National Council of Teachers of Mathematics present a variety of viewpoints. The views expressed or implied in this publication, unless otherwise noted, should not be interpreted as official positions of the Council.

Printed in the United States of America

INTRODUCTION

Using Computer Algebra Systems to Enhance Mathematics Education

OVER the past twenty-five years, calculator and computer tools that enhance mathematics teaching, learning, and problem solving have had a profound influence on school mathematics. Numeric and graphic tools are now in widespread use by teachers of grades K–12 and their students, and growing evidence suggests that such technological tools can expand the scope of curricula and increase students' performance. Computer algebra systems—software that enhances numeric and graphic operations with tools for formal manipulation of symbolic expressions—also has the potential to reshape school mathematics. However, research associated with the influence of computer algebra systems (CAS) on teaching, learning, and the scope of the mathematics curricula for grades K–12 has been limited to a small number of demonstration projects.

In an effort to stimulate thinking and prompt educators to explore and capitalize on the potential of CAS in school mathematics, the National Council of Teachers of Mathematics asked us to prepare a publication that addresses the challenges and opportunities associated with CAS. Our goal is to provide mathematics educators with examples of the best current thinking and recent experiences associated with using CAS tools at the secondary school level. This introduction includes an outline of features common to computer algebra systems and an overview of the chapters that comprise the heart of the book.

What Are Computer Algebra Systems?

As the name suggests, computer algebra systems perform a wide variety of the numeric, graphic, symbolic, and logical operations that form the core components of algebra. Typical computer algebra systems deal with numbers, symbolic expressions, equations, inequalities, functions, vectors, and matrices. Powerful CAS capabilities are now available in several handheld devices and in software that operates on most personal and mainframe computers.

Most computer algebra systems are integrated with tools for producing and manipulating numeric and graphic data, making them multipurpose *computer mathematics systems*. The purpose of this publication is to focus on the formal symbolic operations of CAS that have had much less impact on mathematics teaching and learning than those operations associated with many numeric and graphic tools that are more well known and widely used in mathematics education.

Numbers and Logic in CAS

Typical computer algebra systems assist in mathematical calculation, problem solving, and reasoning by operating on numbers in both exact and approximate forms, by performing common numeric procedures at the request of simple commands, and by using standard logical operations. The following examples illustrate a few of the many kinds of calculations that are made possible by CAS and the corresponding instruction code that is required for each calculation—input and output.

Input	Output
$\sqrt{\dfrac{3}{2}}$	$\dfrac{\sqrt{6}}{2}$
20!	2432902008176640000
Factor (2432902008176640000)	$2^{18} \cdot 3^8 \cdot 5^4 \cdot 7^2 \cdot 11 \cdot 13 \cdot 17 \cdot 19$
$\cos \dfrac{\pi}{6}$	$\dfrac{\sqrt{3}}{2}$
~(a or b)	~a and ~b

Equations and Inequalities

Computer algebra systems are frequently used to solve equations, inequalities, systems of equations, and differential equations. They produce answers in exact or approximate form and deal with expressions of the unknown in terms of literal coefficients. The following are examples:

Input	Output
• Solve $(a*x + b = c, a)$	$a = \dfrac{-(b-c)}{x}$
• Solve $(\sin(x) = \cos(x), x)$	$x = \dfrac{(4k-3)\pi}{4}$
• Csolve $(x^2 + 5x = -8, x)$	$x = -\dfrac{5}{2} + \dfrac{\sqrt{7}}{2}i$ or $x = -\dfrac{5}{2} - \dfrac{\sqrt{7}}{2}i$
• Solve $(3x + 2y = 5$ and $y = x^2, x)$	$x = 1$ and $y = 1$ or $x = -5/2$ and $y = 25/4$
• Solve $(-2x + 5 \leq 2, x)$	$x \geq 3/2$

Equivalent Expressions and Substitution

The CAS routines that facilitate the solution of equations also operate on expressions. Several operations are standard in most CAS, including the following examples:

Input	Output
• Expand $((a-3b)^3)$	$a^3 - 9a^2b + 27ab^2 - 27b^3$
• Factor $(a^3 - 9a^2b + 27ab^2 - 27b^3)$	$(a-3b)^3$
• ComDenom $\dfrac{5}{x-2} + \dfrac{x}{x+1}$	$\dfrac{x^2 + 3x + 5}{x^2 - x - 2}$
• PropFrac $\dfrac{x^2 - 4}{x+3}$	$\dfrac{5}{x+3} + x - 3$
• tExpand $(\cos(3x))$	$\cos(x) - 4(\sin(x))^2 \cos(x)$
• $\sum_{n=1}^{\infty} \left(\dfrac{9}{10}\right)^n$	9
• $x^2 \mid x = \{0,1,2,3,4,5\}$	$\{0,1,4,9,16,25\}$

Functions

When a function is defined in CAS language, routine operations with the previously defined function continue to be available by applying simple instructions. Examples include the following:

Input	Output
• Define $f(x) = 1/x$	Done
• $f(1/3)$	3
• $f(f(x))$	x
• $f(a + 2b)$	$\dfrac{1}{a+2b}$
• $\dfrac{d}{dx}(f(x) + \cos(x), x)$	$^-\sin(x) - \dfrac{1}{x^2}$
• $\int (-\sin(x) - 1/x\char`\^2, x, c)$	$\cos(x) + \dfrac{1}{x} + c$

Widespread use of the amazing calculating tools described above, now available in relatively low-cost handheld devices, has prompted educators to raise fundamental questions about the content objectives of school mathematics courses and the learning opportunities of students enrolled in these courses. Do students, who may some day work in CAS-rich environments, still need to learn the calculation-based skills that have been the staple of traditional curricula? Can educators develop CAS-based explorations that will help students gain a deeper understanding of fundamental mathematical ideas and, at the same time, strengthen their problem-solving skills? How should teachers assess students' knowledge if CAS tools are available to students at all times? Although mathematics educators may have a limited understanding of the issues associated with the preceding questions, the chapters in this book offer them an array of interesting ideas and findings from experiments conducted in classrooms around the world.

Structure of the Book

The papers submitted for publication in this monograph originated in a dozen countries where research is under way regarding the use of CAS in the mathematics classroom. The chapters are presented in four major sections of this book. The first section includes chapters related to ways of thinking about implications of CAS use and ways of making sensible choices for curriculum, teaching, and assessment. The second section includes chapters that describe specific examples of CAS activities used in school mathematics classrooms; this section also includes chapters related to new curricular possibilities. The third section includes reports from specific investigations of student work in CAS environments and also includes a survey of evidence related to the effects of using CAS in mathematics instruction. The final section includes chapters that address the challenges that face educators who assess students' learning when CAS tools are available.

To illustrate how computer algebra systems (CAS) can be used in secondary school mathematics, we have included examples of activities throughout this book. The activities range from fairly ordinary problems to problems that may be too difficult to approach without a CAS. The activities can be used in all high school mathematics courses from first-year algebra to calculus. Although you may want to investigate each activity graphically or numerically, we have chosen to solve them algebraically to illustrate the possibilities of CAS. We hope that you agree that the solutions make the mathematics clear and hide only the symbol manipulation.

In each section of this book, the individual chapters represent a variety of views and experiences associated with CAS tools. The views and experiences are those of the individual authors. We, the members of the editorial panel, have consciously avoided filtering the selections through our own particular perspectives. We believe that the variety of viewpoints presented in this book will make it a stimulating resource for educators who wish to learn more about CAS prospects for the mathematics classroom, to initiate research, or to develop their own classroom projects.

—Editorial Panel
Al Cuoco, Chair
James Fey
Carolyn Kieran
Lin McMullin
Rose Mary Zbiek

PART I

Perspectives for Analyzing CAS Potential: Introduction

COMPUTER algebra systems are here—and here to stay! Computer algebra systems (CAS) will have a far-reaching influence on specifically *what* mathematics is learned and *how* it is learned, so we begin this book by examining the directions in which CAS may lead us. The authors who contributed their expertise and knowledge to this book view the use of CAS in a positive light, but they also recognize the inherent challenges associated with using CAS in mathematics education. Therefore, the following chapters explore the ways that CAS technology affects algebra, the ways that algebra affects CAS technology, and the ways that "CAS enhanced algebra" affects pedagogy.

E. Paul Goldenberg begins by reexamining the various uses for algebra. M. Kathleen Heid summarizes many years of experience and research in mathematics education by discussing theories of mathematics teaching and learning as they relate to using CAS. Bernhard Kutzler presents a framework that leads to a "pedagogically justified" method of using, and not using, technology to achieve teaching goals. William G. McCallum concludes by examining a mathematics problem in detail to illustrate that CAS—rather than provide a magic way to avoid thought—encourage students and teachers to delve more deeply into the mathematics at hand.

—Lin McMullin

Chapter 1

Algebra and Computer Algebra

E. Paul Goldenberg

As I think about the uses of *computer algebra*, I find myself reflecting on the actual uses of algebra—by recalling my first algebra class, approximately one-half century ago.

Seeking a Role for Algebra

When I first began to study algebra as a high school student, I was absolutely sure that I understood the purpose of studying algebra, but I was not sure that I would ever need it for anything other than college admission. Although my high school algebra teachers probably did not use such notation as $f(x)$, they made it obvious to me and other students in the class that the algebraic notation found in such statements as $y = x^2 + 2x - 3$ represented the way that two variables were interrelated. From our viewpoint as students, many of algebra's "uses" came from analyzing equations similar to the one above.

For example, our teachers might have asked us to graph the curve associated with the equation above. We never questioned the purpose of such as task because we took it for granted that graphing was a legitimate goal. We may have accepted the legitimacy of this assignment because scientists and engineers used graphs, or we may have had other reasons for accepting it; either way, we were convinced that, given the task above, algebra would come to our aid! The *form* of the equation $y = x^2 + 2x - 3$ immediately allowed us to identify one special point, the point (0, –3). We knew that the y-axis was the place where $x = 0$, so the curve crossed the y-axis at the point where $x = 0$. The value of y when $x = 0$ could be read directly from the equation; the point (0, –3) was on the curve.

If we were asked to find the roots of the equation above—the points at which the value of y became 0 as the curve crossed the x-axis—we knew that the answer could not be read directly from the equation. We understood that we could try various values of x to discover which one or one(s) yielded a value of 0 for y, but we recognized the need for a more reliable method.

If we were asked to evaluate an expression that was written as a product, we understood that we could use the premise "If the product of two or more numbers is zero, then one or more of the numbers must be zero." But the expression that we were given, $x^2 + 2x - 3$, was expressed as a *sum* rather than as a *product*. If we could rewrite this sum as a product, we knew we would be "all set" and able to solve the equation. Therefore, we learned a technique called *factoring*, a skill that would serve our purposes well. By showing that the statements $y = x^2 + 2x - 3$ and $y = (x + 3)(x - 1)$ were equivalent, we concluded that the value of y must be 0 when, and only when, either $(x + 3)$ or $(x - 1)$ was 0.

We learned about transformations of the graph, but I do not recall actually hearing the term *transformation*. We were taught to recognize the shape of the most "basic" parabola, $y = x^2$, and we understood why the graph of $y = (x - a)^2$ had the same "shape" as the graph of $y = x^2$ but had "shifted" to the right or left, depending on the value of a. I still remember trying to decide in which direction the graph of the second equation moved. The logic was simple enough. In the equation $y = x^2$, y was 0^2, 4^2, 5^2, and 7^2 when x was 0, 4, 5, and 7, respectively; in the equation $y = (x + 1)^2$, y assumed identical values of 0^2, 4^2, 5^2, and 7^2, but in the second equation, the corresponding values of x were – 1, 3, 4, and 6. We came to understand that by adding a constant to the right-hand side of the equation, we simply increased or decreased all the y values by an amount equal to the value of the constant, raising or lowering the entire graph as a consequence. Therefore, if we were asked to graph $y = (x + 1)^2 - 4$, we were able to see that the graph was shaped exactly like the graph of $y = x^2$, but it had shifted to the left by one step and had shifted downward four steps. The point $(-1, -4)$, the vertex of the parabola, could be read directly from this form of the equation.

The constant message was that the *same relationship* between two variables could be expressed by *different forms* of the statement of that relationship. The statements $y = x^2 + 2x - 3$, $y = (x + 3)(x - 1)$, and $y = (x + 1)^2 - 4$ all specified the same relationship between x and y and had the same tables and the same graphs. But each statement revealed information that remained hidden in the other two. Algebra, we learned, provided ways of digging out these hidden meanings from statements by allowing us to convert one form into another. The form $y = (x + 3)(x - 1)$ instantly gave us information about its roots, also referred to as *zeros* or *x-intercepts*, but we had to perform additional "deciphering" to see the y-intercept; we had to do yet something else to find an extreme value—the vertex or minimum. The form $y = x^2 + 2x - 3$ instantly provided us with information about the y-intercept but required us to do some detective work to find the other features. The form $y = (x + 1)^2 - 4$ instantly told us the vertex but hid the intercepts from view until we "dug" them out.

We began to master the techniques of algebra as part of our detective training. To convert an equation such as those above from a form that revealed the y-intercept into a form that revealed the vertex, we needed to see how the picture of the prototype quadratic, x^2, was shifted. First we decided to try to see what expression, when squared, yielded the familiar 0, 1, 4, 9, 16, ... series of square numbers. Where the prototype was x^2, the function we were studying might be built on $(x + 1)^2$ or $(x - 14)^2$. Next we needed to know to what extent these square numbers were camouflaged by the addition of a constant—in other words, how much of the picture was "shoved" up or down.

We knew that we needed to become very familiar with such identities as $(x + a)^2 = x^2 + 2ax + a^2$ to become proficient at the skill of completing the square. I remember squaring long lists of binomials similar to $(x + 7)$, $(x - 12)$, and $(x - a)$; I also remember factoring trinomials in the form of $16x^2 - 24x + 9$. When we were finally able to factor perfect square trinomials accurately, we looked at the equation $y = x^2 + 2x - 3$ and thought silently, "The trinomial $x^2 + 2x - 3$ would be a perfect square trinomial *if only* the constant term were the number 1 instead of –3. Well, we will simply *let* it be 1, and then repair the damage later." The new statement $y = x^2 + 2x + 1 - 4$ was simply another *form* of the original statement, but it took us one step closer to rewriting the expression in yet another form, $y = (x + 1)^2 - 4$, thus revealing something that we wanted to know but could not see when the equation was expressed in its original form, $y = x^2 + 2x - 3$.

Algebra allowed us to find numbers that simultaneously satisfied *two* equations. Once again, we knew that we must change the form in which true statements are made. If we were told that $y = 3x + 5$ and $y = 5x - 3$, then surely $3x + 5$ must equal $5x - 3$ because "things equal to the same thing are equal to each other." On the basis of our "mathematically oversimplified" assumption, we knew that $3x + 5$ must equal $5x - 3$. We also knew that our new statement, $3x + 5 = 5x - 3$, could be changed in form to show that $x = 4$; we could then substitute the value 4 for x in either of the original statements to show that $y = 17$.

Algebra seemed to be all about *form*. As we solved algebra problems, we often started with something that we knew, devised a way to express it, and then changed the form of that expression to reveal something that we did not originally know. We were assigned many word problems, problems about ages, trains, and money, the simplest of which did not seem to require any algebra unless the numbers were cumbersome. We also worked with equations of a line in various forms, such as point-point, point-slope, slope-intercept, intercept-intercept—more examples that reinforced the idea that algebra was about form and changing forms. If we could find a way of changing the form without altering the meaning, we could reveal hidden truths!

Of course, some students legitimately griped, even then, that they did not care about these hidden truths and could not imagine ever needing to ferret them out. Often enough, the nature or quantity of the exercises that these students were assigned numbed rather than stimulated their minds, leaving some "mathematical truths" as well hidden at the end of the course as they were at the beginning. Aside from the fact that some of us loved algebra and some of us did not—perspectives that were often based on its personal appeal and utility to individual students—the *purpose* of algebra as an intellectual discipline was fairly obvious. Under the guidance of a good teacher, we understood the purpose for learning individual techniques. The process of *expanding* changed products to sums; *factoring* changed sums to products; *simplifying*, well, simplified. Sometimes, we even "complicated" a step that we made en route to completing the square, as in the example of the equation $y = x^2 + 2x + 1 - 4$. Form was the essential element in algebra.

Any discussion of educational policy and practice that does not deal with the political side of an issue is incomplete. Algebra, when taught as a general education course, must benefit those students who will pursue careers fields that require advanced mathematics and also those who will not pursue such careers, the latter group historically a *much* larger group than the first. Very few people will need to factor trinomials when they are thirty-five years old, not even those who learned to factor them in high school so that they could be admitted to college. Yet the way in which algebra is taught should not discourage people from pursuing mathematics-related careers before they have had an opportunity to develop their interests, tastes, and abilities.

We read that success in algebra correlates with such real-life successes as better pay and better marriages—surely not a result of knowing or not knowing how to factor! Right or wrong, algebra remains a gateway to opportunity. Therefore, a lack of access to, or success in, algebra has sweeping negative consequences unrelated to whether a person ever experiences a personal need to find the vertex of a parabola. However, access to algebra is not equal, and success or failure is predetermined in part by previous school experiences over which the student has no control. Algebra has social consequences, as do many other school subjects, and has been part of the filtering process that perpetuates separation by class and privilege. Although it *does* serve this purpose—and one can cogently argue that despite calls for equity, the entire system of education is well designed to serve this purpose—we cannot assume that anyone planned it this way. To make such an assumption would require too elaborate a conspiracy theory. The effect is destructive, and people, not randomness, are to be blamed; but the cause is negligence or disregard rather than an imaginably conscious plan. I am sure that no one perceives the purpose of algebra to be anything other than purely mathematical in nature.

If educators are to teach the *details* of algebra—thus protecting the interests of students who need to learn details without wasting the time of those who will never need to understand the details of algebra—they must focus their attention on a "loftier" goal. Perhaps algebra should focus on the idea of detective work or on the notion that "changing form without changing meaning" can help make obvious what would otherwise remain obscure. I will present more information about the role of form and meaning later in this chapter because they are part of our daily lives; the role of form is exemplified in the work of automotive diagnosticians, medical diagnosticians, and investigative journalists as they reveal the unseen and express themselves in a variety of ways.

Fitting Technology into This Picture

If the subject at hand continues to be *form*, along with the power and facility to recognize the structure of a computation and the ability to reflect a different structure, then computer algebra systems present a daunting challenge. Computer algebra systems (CAS) do, with no effort, what we previously thought we wanted the *students* to do!

I first became aware of this conflict at a time when symbolic manipulation was still so expensive that it was not a serious player in the high school education arena, but the shock was enormous for teachers of college calculus. Before the existence of a calculator capable of performing symbolic differentiation and integration, teachers could not underplay mechanics and attend only to big ideas or important applications, no matter how much they might have wanted to do so. Such techniques as integration had to be mastered, and because they were difficult skills for students to learn, an enormous amount of course time was spent on teaching them. Suddenly, students were able to buy a little box that did all the procedures for them and that cost less than it cost them to learn how to do it themselves, As a result, the primary focus of many calculus courses was questioned, and instructors had to re-evaluate what parts of their courses were actually important. Perhaps learning some of the techniques that the calculator now performed would continue to be important, but educators could no longer argue that "getting the answer" was part of that reason for asking students to learn these techniques.

The problem facing high school algebra is a bit worse. Not only must we, as educators, decide which algebraic manipulation techniques to relegate to the little black box and which calculus and arithmetic techniques to continue to teach to students, we must also rethink the purpose of algebra as a result of the influence of modern tools and educational emphasis. Graphing technology—the computer and then the ubiquitous graphing calculator—delivered a major blow to algebra. If algebra is helping students find roots of equations, slopes, tangents, intercepts, maxima, minima, solutions to systems of equations in two variables, or any of the other numerical and application-related answers, then quick and easy access to a graph is all our students need. Students no longer need to change the *form* of a mathematical statement, because any form of an equation will graph as easily as any other. Students not only do not need to perform algebraic manipulations by hand, they do not need CAS to do the manipulations for them either!

Another blow against the algebra of yore is the emphasis on "applications outside mathematics." If one needs to know how long to make the diagonal brace for a 3×3 square wood frame, then $3\sqrt{2}$ is not only an example of unnecessary precision, it is incorrect! From a notational standpoint, $3\sqrt{2}$ is more compact than 4.24264069 or 3 × 1.41421356, and πr^2 is a more sensible way to express the formula for the area of a circle than the "over precise" expression $3.14159265r^2$. Similarly, the quadratic formula supplies information that cannot be found by examining roots on a graphing calculator, but the information is needed only in mathematical situations rather than real-life situations. In fact, the symbolic forms and precise manipulation of irrational numbers are of little use in applications outside mathematics, except when used as shorthand or for clarification.

I believe that wide access to graphing technology and the current increased emphasis on applications are not entirely unrelated. In hindsight, I think that graphing technology was first viewed by many educators as an unimportant example of the "multiple representations" that would help students gain a better sense for algebraic symbols—another route to algebra. For example, Yerushalmy (1999) uses the graphical approach to help students make sense of arithmetic operations on functions, function composition, and the various algebraic manipulations used to solve equations. Interestingly, complex numbers, which receive only a fraction of the attention that is generally paid to irrationals, are used in many application situations. Yet over a ten-year period, graphs became what algebra was *about* rather than simply one of the tools available to educators to help students understand algebra. Part of this transition may well have been the result of the success of the graphical approach, but that success was abetted by at least one other factor.

To use graphing tools *at all*, the functions must be expressible in an algebra language that the tools understand. As graphing tools emerged, such a requirement seriously restricted the domain. Furthermore, for graphs to be used to find numerical answers—in interest-related problems, for example—the graphs must be continuous and well behaved by nature. In spite of both the limitations described above, the graphing tools are well suited for many topics, such as simple physics or economics, because many of the applications associated with these subjects are modeled with continuous "simple" functions. Thus, the tools favor one type of mathematical problem domain over another, elevating the role of the extra-mathematical-modeling role above that of the structure-of-calculation role. By using suitable manipulations in that favored domain, educators find that the tools have become a substitute for algebra as a means to solve equations. At the same time, the tools have left algebra intact as a language in which to express equations.

Is Algebra Dead?

Calls for changes to high school algebra, in response to the newly available computing technology, date back as far as 1985. An NCTM special conference on the impact of computing technology on school mathematics (Corbitt 1985, p. 246) stated,

> The skill objectives of algebra must be reassessed to identify those procedures more easily done by computer ... or calculator... The properties of elementary functions are still important for modeling quantitative relationships, but proficiency in many familiar computational processes is of little value.

The sketchy statement above was made during the time that computational technology in schools was still crude by today's standards, but the sentiment expressed remains quite current. The argument below will probably seem familiar.

If algebra is useful only for finding roots of equations, slopes, tangents, intercepts, maxima, minima, or solutions to systems of equations in two variables or for finding other numerical application-related answers, then it has been rendered totally obsolete by cheap, handheld graphing calculators—dead—not worth valuable school time that might instead be devoted to art, music, Shakespeare, or science. For all the numerical answers listed above, we do not need exact solutions, as if exact solutions can be found much of the time anyway! We need no more algebra than is required for encoding real-world situations in a way that our black box can understand. At the push of a button, the box then performs calculations with precision far beyond what is required for most practical purposes. We do not need to use the distributive law; we do not need to factor; we do not need to complete the square; we do not need to use the quadratic formula. We *do* need to understand how to encode situations into suitable symbols for the machine. We *do* need to know what question to ask the machine to solve, including what statistical tool to invoke and what part of a graph to inspect closely. We *do* need some plausibility checks, such as estimation strategies, that will help us avoid mistakes resulting from pressing the calculator keys incorrectly; we also need plausibility checks that will help us avoid unfeasible answers to real-life problems, such as making reservations for 2.8409091 school buses. Of course, we also need to know what the various new computational tools are, how to use them, and how to choose sensibly among them when we are trying to solve a problem.

The conclusions in the previous paragraph appear to be accepted by very many people, because one encounters them frequently. If the argument seems unassailable, note carefully that it rests entirely on its first word, *if*, as it concerns the computational uses of algebra.

The process of finding of roots, intercepts, extrema, and other similar features does not require more than a small box and knowledge of its use, but determining such information is *not* the only use for algebra.

Hidden Truths Revealed

In the past, school algebra served many masters. It helped people model and solve practical problems in science, engineering, business, and other real-life occupations; it also helped describe the structure of calculations. Some of those masters are now gone. The problems for which algebra was used outside of the realm of academic mathematics typically required numerical answers, so algebra was primarily used by people to encode a real-world situation in formal mathematical language; algebra was also used to manipulate the encoding—usually an equation—until it revealed its hidden meaning, possibly a solution, an extremum, or a rate. Algebra really is *not* needed to reveal these hidden truths, although it is still needed by people who design the little black boxes that we call "technology."

But algebra always played another more subtle role in opening up various black boxes, including the ones we called *patterns*. Since elementary school, we knew, for example, that the sum of any two odd numbers was always even, but why? That reason remained unexplained until algebra opened the black box. We did not need to know the explanation, and few of us asked; but most students, including those who did not generally like algebra, thought it was "cool" to know what underlay the patterns we had seen. Bastable and Schifter (in press) report that children do regularly express curiosity about the mechanisms behind the patterns they see, genuine interest in working out those mechanisms, and an algebraic way of thinking without the conventional notation. In a similar way, we saw that sums like 1+ 3, 1+ 3 + 5, 1+ 3 + 5 + 7, and so on, were always square numbers, but why? The idea that "the nth square number plus the $(n + 1)$st odd number is the next square number" is encoded nicely in the algebraic identity $n^2 + (2n + 1) = (n + 1)^2$. So $1^2 + 3 = 2^2$, and $2^2 + 5 = 3^2$, and $3^2 + 7 = 4^2$, and so on. Do we need to know this fact or its explanation? The answer is not clear, but if the curriculum contains the preceding fact, even simply as a "cool" pattern, then algebra helps transform it from *Yet Another Unexplained Mathematical Phenomenon* (YAUMP) to an understandable and nonarbitrary part of a coherent system of ideas. YAUMPs "disempower" students. Yet taught well and given the correct emphasis, algebra might be the antidote for the common complaint that educators are teaching a miscellany of disconnected procedures to be learned by rote, a manner that is antithetical to students' understanding. Algebra could *serve* understanding by solving mysteries and showing students that effects have causes.

A new emphasis must be implemented because some algebra skills that are no longer needed for *finding* answers still remain essential for *understanding* answers or the methods by which the answers are found. Consider, for example, the division algorithm. Students no longer needed to find answers to division problems; *nobody* who does much division ever uses by-hand division, even if the calculator batteries die. Yet division may be the best avenue for helping students understand several mathematical ideas, both, algebraic and nonalgebraic, such as the frequently taught fact that rational numbers can be expressed as repeating or terminating decimals. We might reasonably argue that this fact is not important enough to teach, but most curricula deem it otherwise; and *if* it is to be learned, it should not be YAUMP. One way to understand and prove that rational decimal expansions terminate or repeat is to examine the division algorithm closely and watch the way that the digits are generated.

In extramathematical domains that the graphing calculator favors, we not only need limited symbolic manipulation skill ourselves but also need few of the manipulations to be performed by machines. In the domains described above, the hidden information is generally numerical but the technology enables us to arrive at numerical solutions without attending to form.

Algebra's applications inside mathematics remain intact because they focus less on numeric solution than on the expression, explanation, proof, investigation, and extension of mathematical phenomena. For example, the graphing calculator cannot show that each odd number is the difference of two perfect squares or that the product of two odd numbers is odd. The symbolic calculator can help people perform the manipulations, but even that aid is of little value unless they have figured out how to formulate the problem clearly and know which patterns to seek, an understanding that comes from studying the structures itself. On the one hand, students can easily see that 9 divides 99999 or 9999999 or 99. By using a symbolic calculator, students can also easily show a polynomial analogue, such as that $(x - 1)$ divides $(x^n - 1)$ for many values of n. On the other hand, proving that $(x - 1)$ divides $(x^n - 1)$ for all values of n requires students to see how the calculation works algebraically and argue, through induction or informally, from the structure of the calculation.

Are intramathematical purposes, such as showing that the product of odd numbers is odd, worth our attention and our students' time? The answer to the preceding question is a matter of opinion, one on which reasonable people often disagree. I personally believe that students need an understanding of how mathematics works, not merely how it applies, so that mathematics does not become simply another incomprehensible tool in our endeavor to teach the concepts of number, symbol, spatial interpretation, and data "sense." Without an understanding of mechanism, no real sense can be made.

In fact, a focus on the pragmatic utility of mathematical results may work quite generally against the development of mathematical sensibility. If the value of a particular mathematical result rests in its utility, students do not need to understand why or how it works or pursue the thinking to prove that it works. This premise supports an epistemology of facts, not reason, in spite of our intent to do otherwise. Similarly, technology permits a "how to" and "button pressing" view of the discipline, already too deeply ingrained in popular culture.

Possible and Impossible Roles for CAS

Many attempts to decide what does and does not constitute essential algebraic knowledge have ended without broad agreement, but not because of a lack of good thinking or good intentions. The community of teachers, mathematicians, parents, students, and others with legitimate interests and relevant and informed opinions is a complex entity and not of a single mind. The environment in which the decision to define essential algebraic knowledge must be made is an environment of changing technology, demands, tastes, communities, and teaching staffs; conditions vary in ways that may render a single policy recommendation inappropriate.

Because outside sources have not reached consensus regarding the issue of essential knowledge, and will probably not do so in the near future, you need to think this out in relation to your state or local standards, your community, and your own students. Here are two ideas with which to agree or disagree as you do your own thinking. I state them as conclusions, for they are conclusions, and I explain the reasons that make these conclusions valid to me. But you must remember that they are my conclusions, at least for now and for the students I know, and these conclusions may or may not be equally valid for circumstances as you see them.

Abstract form remains important

In a fascinating presentation at NCTM's 2001 Annual meeting, presenter John F. Mahoney asked the question, Which is the interesting result,

$$\int_0^1 \frac{x^4(1-x)^4}{1+x^2}dx = 0.00126449$$

or

$$\int_0^1 \frac{x^4(1-x)^4}{1+x^2}dx = \frac{22}{7} - \pi \ ?$$

Similarly, if $a_1 = 1$, which is the interesting way to characterize the lengths of segments a_2 through a_7 in **figure 1.1**? As 1.41421, 1.73205, 2, 2.23607, 2.44949, and 2.64575 or as $\sqrt{2}, \sqrt{3}, \sqrt{4}, \sqrt{5}, \sqrt{6}$, and $\sqrt{7}$?

Figure 1.1

Can you characterize the lengths of segments a_2 through a_7 in an interesting way?

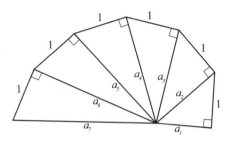

In both cases, the results are interesting only when the structure is apparent. In fact, even the elementary school pattern behind $1 + 3 + 5 + 7 = 16$ becomes more interesting when one sees the structure: $1 + 3 + 5 + 7 = 4^2$. Abstract form remains important, even in students' eyes, as illustrated by the following slight variant on the example of $1 + 3 + 5 + 7 = 4^2$. Given the set of computations shown in column (a) of the chart below, young children almost invariably find columns (c) or (d) more interesting as a type of answer than column (b).

a	b	c	d
2 + 4	6	2 × 3	2 × 2 + 2
2 + 4 + 6	12	3 × 4	3 × 3 + 3
2 + 4 + 6 + 8	20	4 × 5	4 × 4 + 4
2 + 4 + 6 + 8 + 10	30	5 × 6	5 × 5 + 5

The numbers in column (b) are simply numbers. *Any* numerical computation will produce *some* number! The numbers become interesting only if they happen to be old friends showing up in new places. In all three school subjects—calculus, geometry, and elementary arithmetic—the more interesting result is the one that reveals some structure or idea by its form or makes some connection that is unexpected.

In programming classes, a great deal of attention is spent on form. The explanation "clean code is easier to debug" is usually given to students, but underlying this explanation is the much bigger idea that "clean code is *conceptually* clear." In summary, *form* reveals the underlying idea that messy code obscures, even if it happens to work correctly. Even in very simple programs, such as the two Logo procedures shown below for drawing a square, the differences in form are more than a cosmetic matter.

Algebra and Computer Algebra

Procedure 1

   to square :d

   forward :d right 90 fd :d rt 90 fd :d rt 90 fd :d rt 90

   end

Procedure 2

   to square2 :distance

   repeat 4 [forward :distance right 90]

   end

Each of the two procedures shown above moves a drawing cursor "forward" a specified distance in the direction it was facing originally, then changes its heading by turning it to the right 90 degrees, and then moves it forward the same distance in the new direction until exactly four moves and four turns have been completed. The second procedure is not merely fewer keystrokes but is a "simplified" form of the first procedure that is analogous to algebraic "simplification," and, in this instance, similar to factoring. The second procedure reveals structure—the fourness of a square—that is not readily seen in the first procedure.

The manipulative aspects of algebra remain important

After debating the importance of arithmetic skills for elementary school students in an era of inexpensive and ubiquitous arithmetic calculators, many educators—in all their "current wisdom"—have concluded that elementary school students still need to know *what* the arithmetic does for them and *how* to perform it. Many educators do not think that the use of extensive drill with large calculations, for the purpose of developing efficiency and accuracy, is good use of students' learning time. To produce approximate calculations—whether to interpret calculator results or to avoid the need for them—students need to know both facts *and* procedures and demonstrate moderate proficiency in using them.

Elementary school students need to acquire the skills described above so they can understand the black boxes of computation and learn to recognize patterns, but the convenience of having quick access to many building blocks is important. The more basic facts that students simply *know* and, therefore, do not need to work out, the more fluent they will be in calculating. On the one hand, knowing a large store of facts helps students recognize and make use of patterns. For example, students are more likely to become curious about the pattern 1, 1 + 3, 1 + 3 + 5, and so on, if the set {1, 4, 9, 16 . . . } is familiar to them and they are not surprised by the contextual shifts from sums to products. On the other hand, "more is better" *only* to a certain point. Although memorizing all the products of pairs of single-digit numbers may be useful for students, it does not follow that memorizing all the products of pairs of double-digit numbers is better yet for them. The learning cost is too high.

The same arguments apply to algebra. Algebra students must choose from among learning the arithmetic operations, factoring, expanding, simplifying, and *complicating*. Even if students rely entirely on a calculator to perform these algebraic manipulations, they must know which procedure to choose, its purpose, when it can apply, and when it is *likely* to produce the desired effect—decisions that require a strong sense of the nature of the manipulation. The possibility that students could acquire a sense of the nature of manipulation from CAS use alone is only barely imaginable, but the plausibility that they need to perform operations themselves to understand the nature of the operations is more believable. As with arithmetic calculations, the level of discipline that students developed in the past to tackle great computational complexity may no longer be worth the time spent. Massive computations probably should be relegated to tools that have care and discipline etched into their circuits. Students need to understand the principles behind such computations and the way they are performed; they should also become competent at performing computations in problems that require several steps.

Peter Braunfeld of the University of Illinois at Urbana-Champaign suggests the existence of an algebraic analogue to the basic facts of arithmetic—algebraic computations that are so common and so useful that they should be considered basic facts. As in arithmetic, the process of creating such a list involves finding a moderate position on which everyone can agree. The following very brief and tentative list is far too new to have withstood a test of "moderacy." It is best thought of as a discussion-starter—a proposed first draft list of "basic facts" and problems that students need to find "easy" but should be able to derive and understand.

1. Some "basic facts":

   a. $(a + b)^2 = a^2 + 2ab + b^2$ and $(a - b)^2 = a^2 - 2ab + b^2$

   b. $(a + b)(a - b) = a^2 - b^2$

   c. $x^{-n} = \dfrac{1}{x^n}$ and $x^n \cdot x^{-n} = 1$ $(x \neq 0)$

   d. If $ax^2 + bx + c = 0$, then $x = \dfrac{-b \pm \sqrt{b^2 - 4ac}}{2a}$ (the quadratic formula).

e. If $a > 0$, then a has two real square roots: $\sqrt{a}$ and $-\sqrt{a}$, and if $a < 0$, then a has no real square roots.
2. The distributive law and notation:

   a. $n\left(a + b + 2xy + 77p^{3+\sqrt{q}}\right)$

   b. x times a polynomial in x, e.g., $x\left(2x^3 - 3x^2 + 77x^{-2+\sqrt{q}}\right)$

   c. $2(2a^3 - 3ab^2 + 7b + 7)$ and $-3(2a^3 - 3ab^2 + 7b + 7)$

   d. $a(2a^3 - 3ab^2 + 7a + 2b + 7)$

3. Common nonlinearities:

   a. $(a + b)^n \uparrow a^n + b^n$

   b. $|a + b|$ does not, in general, equal $|a| + |b|$.

   Rather, $|a + b| < |a| + |b|$ (the triangle inequality).

4. The different behaviors of sums and products:

   a. $\dfrac{c}{a+b} \neq \dfrac{c}{a} + \dfrac{c}{b}$. But $\dfrac{a+b}{c} = \dfrac{a}{c} + \dfrac{b}{c}$.

   b. $\sqrt{a+b} \neq \sqrt{a} + \sqrt{b}$, but $\sqrt{a \cdot b} = \sqrt{a} \cdot \sqrt{b}$.

   c. Knowing when to factor and when to expand

5. Given problems like the following on paper, students should be able to solve them mentally without writing explicit operations or performing substeps.

   a. Solve $2x + 5 = 10$ or $3x + 5 = 10 + x$ or $x^2 + x + 5 = x^2 + 10$.

   b. Simplify $a(b + c) - ab$.

   c. Simplify $a\dfrac{a}{\sqrt{2}} - \dfrac{a}{2\sqrt{2}}$.

Empowerment requires control

So far, I have focused almost exclusively on the importance of learning *mathematical* ideas, skills, and knowledge. I have also discussed ways that technology might aid learning or impede that learning. From the perspective of mathematical ideas, skills, and knowledge, mathematics is the goal and technology may be the means. I also think that at times the competent use of a technology can become a legitimate goal in itself, even when the name and purpose of the course continues to be *mathematics*. The alternative perspective mentioned above is easier to understand if placed in the context of learning a language.

For example, teachers of writing rebelled long ago against a tradition of teaching technical details devoid of communicative purpose. They believed that focusing solely on the mechanics was stultifying to students and discouraged expression, which was the true purpose of writing. This realization was important and correct, yet in some situations the solution to this problem resulted in a pendulum swing that worked "its own mischief." To have something to say but be unable to say it because of the need to pay so much attention to minutia is a "defeating" experience. To have something to say but be unable to communicate it adequately for lack of skill is equally defeating.

For example, I feel slightly stupid when I speak French. I feel this way not because I cannot hold a conversation—I *can*—but because what I am able to say is not as sophisticated as what I am able to think and because I cannot be confident that I understand the subtle nuances in what others say. I suspect that very smart people can feel stupid in mathematics for quite similar reasons. We, who frequently champion the use of technology, point out that it can help students think the "big" thoughts before they have learned the language; the traditional language of the algebraic machinery, among other factors, leaves them apparently unable to function mathematically.

As evidence, educators refer to the many "big ideas" of mathematics, for example, the intermediate value theorem and the fundamental theorem of calculus—theorems that are quite natural and can be grasped and honed by students who are relatively young if the language of algebra does not get in the way. Technological alternatives to traditional algebra—from computer algebra systems and programming languages to graphing calculators, from spreadsheets to elaborate special-purpose software—have been proposed as legitimate new access routes to these important ideas.

When, and to what extent, should we shift from the traditional machinery to the new technologies? The dilemma is reminiscent of the debate surrounding bilingual education, with wisdom evidenced on both sides of the debate. Proponents of one side of the bilingual education issue believe that students' intellectual development should not be impeded by their lack of facility with a particular language when another language would allow them to flourish: if English is a weak link, then educators should encourage at least some learning to proceed in a more appropriate language. The views expressed by people on the opposing side of the bilingual issue are also important to hear. They believe that students should not be denied full fluency with the language or languages they need for discourse in their domain. In the bilingual education debate in the United States, the *language* that students need is *English*. Similarly, in the field of mathematics education, if software can provide students with access to important ideas before the machinery of algebra is fully developed for them, then educators should use it; the *language* that students need in mathematics education is *algebra*.

Students were not masters of the *old* tools—specifically, algebra. The lack of competence hampered them and left them without power. We are not doing students a favor if we give them *new* tools that they do not master either. However, with a limited set of the most powerful tools, we might treat the technology as a goal as well as a technique, just as algebra is a goal as well as a technique. As long as our approaches to using CAS are not intellectually empty, we can provide students with its own worthy organization of the content. We should then find systematic ways to develop competence with the technology as students move from grade to grade, so that high school students can use the tools easily and appropriately to solve nonroutine problems that match their intellectual and mathematical development.

Although algebra is *more* than a language, I find the parallels between learning algebra and learning a language to be very strong. Algebra is of little or no intellectual value to those who can use it only to solve preformulated routine exercises, but it is of great intellectual value to those for whom it is a fluent and expressive medium. The same, I believe, is true for CAS or any other electronic tool. If programming, spreadsheets, interactive geometry software, or CAS are to be of intellectual value to students, they must know enough to use the tool comfortably to represent and manipulate novel mathematical problems. Such tasks involve *understanding the mathematical domain* well enough to know what to ask of the tool and *understanding the tool* well enough to know its capabilities and how to get it to work. Educators who developed the current curricula and frameworks talk about using the new technologies, but they do not seem totally committed to equipping students with such facility. Students may see simple computer programs in their textbooks, copy the programs, and run the programs on the computer, but they may not be able to construct a similar program that they need to represent an algorithm they are studying. Using geometry software, students may not be able to build their own models of interesting mathematical phenomena without having the steps or button-presses specified for them. Beyond the most elementary constructions, we teachers or curriculum developers are the makers of the models that we think students should see. For example, students using CAS may be told to "use factoring to show…" or may be instructed to use spreadsheets to enter tables and create graphs, but they typically learn very little of the incredible power inherent in these tools.

Algebra can be approached two different ways: either as a long list of tricks—analogous to pull-down menus and buttons to push—or as a collection of important ideas expressed in an orderly and relatively consistent way. The first perspective relegates it to a kind of "overhead," having little merit in its own right but being a necessary skill to acquire before moving on to the "real stuff," such as calculus. Similarly, geometry software can be learned entirely as overhead—not part of the mathematics but part of the tool acquisition en route to worthy mathematics—or as an embodiment of a valuable set of ideas.

Peter Braunfeld, in a commentary on a precursor to this chapter, presented the following notion that the very existence of a button on the software may be perceived by students as evidence that something of "real" importance lies there:

> I think that when children are asked to use technology to solve problems, that in itself can make the problem seem more important to them. For example, the mere fact that buttons for certain functions and operations appear on a calculator endows these functions and operations with a certain "gravitas." The kids know that making calculators is a big business, that real people use them in the course of their important work, so why would the manufacturers have put these buttons on the machines if they weren't important functions?
>
> For example, if no one cared about **ln**, that button surely wouldn't appear on the calculator.

We would be perverse to allow technology to drive mathematical content. The fact that we can do new things does not necessarily mean that we should do them, and the fact that we can have a box that will do old things for us does not necessarily mean that we should no longer know how to do them for ourselves. Applying Braunfeld's ideas to the world of CAS, we might refer to the **factor()** button and ask ourselves why anyone would ever want it to use it, what kinds of objects it applies to, and what kinds of objects it returns. Students who know how to factor a number like 49 into $7 \cdot 7$ may not understand why computer algebra systems will not factor x^2 into $x \cdot x$. Similarly, when students factor $ab - ac$ to be $a(b - c)$, they separate a from b, but when they are taught to factor $abx - aby$ into $ab(x - y)$, they may not see why they should not also separate a from b when factoring $abx - aby$ into $ab(x - y)$. In fact, why should factoring $abx - aby$ give $ab(x - y)$ and not, for example,

$$ab^2\left(\frac{x}{b}-\frac{y}{b}\right)?$$

In all these examples, the answers are not idiosyncrasies of the technology or even the procedural details of hand factoring; they are the very purpose of this algebraic operation and the reasons we might need it.

Students also need to see beyond what the CAS tool will do. For example, given the directions "If x and y are divisible by b, show that $(x-y)$ is also divisible by b," one student used the notion of "factoring" $(x-y)$ into b. The student translated "n is divisible by b" to

"$\frac{n}{b}$ is an integer"

and wrote, "If

$$\frac{x}{b} \text{ and } \frac{y}{b}$$

are integers, then

$$\left(\frac{x}{b}-\frac{y}{b}\right)$$

is, too. Because x and y are both divisible by b we can factor $(x-y)$ to get

$$b\left(\frac{x}{b}-\frac{y}{b}\right)."$$

Although the statement above represents a loose and unconventional use of terminology, it is an example of perfect use of the concept of factoring and its connection with divisibility.

Posing different problems as a result of CAS

Many decisions must be made regarding exactly what algebra might help students learn. One such decision involves deciding whether such skills as factoring and expanding should remain among the subgoals of algebra—"subplots" that are essential elements in the development of the "algebra story"—rather than serve as the ultimate plot of the algebra story. Taking the position that the skills outlined above do remain important, I want to find ways for students to practice the skills. Such exercises afford little practice if they are worked by CAS-using students who pay little attention to process. Therefore, I can advocate either banning the use of a CAS or coming up with a new kind of activity for which CAS will be useful—without subverting the purpose of the exercise. For example, traditional treatments of factoring first explain how to factor, then provide some worked-out problems, and conclude with a collection of exercises that are essentially "random" in sequencing and order; students appear to receive roughly the equivalent benefits from doing only the even exercises, doing only the odd exercises, or doing the exercises out of order. The reason that the configurations just described are equivalent is that the practice lies entirely in the doing of problems; the problem-set directions do not generally require students to examine answers for *structure* or other information that the answers might reveal. If, in a problem set, students are asked to analyze some of their answers, then how the answers were attained—whether by hand, by CAS, or copied from someone else's work—may not matter. Below is a very abbreviated sketch of such a problem-set; problem 1 can be done either by hand or by CAS; problem 2 requires an analysis of the results; problem 3, which cannot be done by CAS, requires that students understand the structure of the answers to problem 1.

1. Expand these squares to complete the statements.

   a. $(x + 6)^2 =$
   b. $(x - 3)^2 =$
   c. $(x - 10)^2 =$
   d. $(x + 2\frac{1}{2})^2 =$
   e. $(x + \sqrt{7})^2 =$
   f. $(x + a)^2 =$
   g. $(x - a)^2 =$

2. Describe a pattern in the way that your answers to problem 1 relate to the expressions you were told to expand.

3. Use that pattern to figure out what value a and b must have to make the following statements true.

   a. $(x + a)^2 = x^2 + 6x + b$
   b. $(x + a)^2 = x^2 - 6x + b$
   c. $(x + a)^2 = x^2 + 9x + b$
   d. $(x + a)^2 = x^2 + bx + 9$
   e. $(x + a)^2 = x^2 + bx + b$
   f. $(x + a)^2 = x^2 - bx + 81$

For learners, a specific mathematics problem is more than just a problem; it is an illustration of something more generalized—a real object of their learning. The "answer" at the end of that problem is either a check on students' handling of the ideas or techniques or a piece of data contributing to a pattern they must then notice. Such a pattern may help students understand a structure, process, generalization, way of thinking, or other organizing idea, and such understanding is the true goal. If technology provides answers so quickly that the process and intermediate results are suppressed, students may never see the structure that lies behind the computations and thus may miss the important part. One example of such a scenario is the production of 3s, one by one, in the decimal expansion of 4/3. Viewed on a calculator, 1.3333333 is simply an answer. Performed by hand, in the right instructional context, the decimal expansion described above has a "rhythm" to the calculation that may illustrate something about the infinite process and the reason it never ends. Another example of the validity of the foregoing premise is $(x-1)(x^7 + x^6 + x^5 + x^4 + x^3 + x^2 + x + 1)$ as it collapses to produce $(x^8 - 1)$. The process of simply seeing the answer to the preceding problem does not offer students much insight into the process involved, but seeing the *intermediate* results gives them in-depth insight.

If, for example, a computer algebra system is helpful in gathering the data and producing a set of polynomials of the form $(a + b)^n$ in which the binomial coefficients are examined, it may reduce the distraction and error that accompany complex calculations. One reasonable way to decide when to use technology to reduce the work involved in a computation is to ask oneself if the given computation is a distracting step in the midst of a process that is being studied or if it is the process to be studied. The sequence of problems shown in the preceding numbered examples are designed to help students recognize the resulting pattern when they *factor* perfect square trinomials. For that reason, problem 1 above need not be done by hand. If the object of the lesson had been to teach students to square binomials, an obvious benefit would have been realized from students' doing exercises like problem 1 by hand.

Conclusion

Technology truly invites educators to examine the content and techniques associated with traditional mathematics and decide what can be "jettisoned" to make room for new ideas and approaches. On this subject, I find myself "of two minds." On the one hand, the idea that we *should* "throw over" certain content that was previously considered essential seems obvious; to some extent, we do so now. For example, learning to extract square roots is pointless when calculators render effortless a task that otherwise is hard, distracting, and of relatively little intellectual merit. Students should be allowed to use the calculator and reassign the time to a more useful purpose than they previously spent on learning to extract roots. On the other hand, our poor history of success in teaching algebra to students and the ready availability of CAS have provoked educators to question whether algebra, like the extraction of square roots, may be in the process of becoming obsolete.

To envision such a fate for algebra is difficult for me. Perhaps algebra *will* be obsolete someday, but until that time we must continue to teach it.

We do have reasons for drawing on such tools as CAS in the service of such teaching. Levasseur and Cuoco (this volume) give examples of the technologies that are drawing people toward mathematics. With the existence of such powerful and versatile tools, it may be as great a gift for students to acquire the same kind of fluency with the new tools that we had always hoped they would develop with mental and paper-and-pencil skills. More to learn? Perhaps, but learning may also be better and faster. It will take thought and new curricular approaches, and it will take time!

Acknowledgment

This chapter was supported, in part, by NSF grant ESI-9731244, Connecting with Mathematics. The opinions expressed are those of the author and not necessarily those of the Foundation.

REFERENCES

Bastable, Virginia, and Deborah Schifter. "Classroom Stories: Examples of Elementary Students Engaged in Early Algebra." In *Employing Children's Natural Powers to Build Algebraic Reasoning in the Content of Elementary Mathematics*, edited by James Kaput. In press.

Corbitt, Mary Kay, editor. "The Impact of Computing Technology on School Mathematics: Report of an NCTM Conference." NCTM Conference Steering Committee. *Mathematics Teacher* 78(4) (April 1985): 243–50.

Goldenberg, E. P. "Habits of Mind as an Organizer for the Curriculum." *Journal of Education* (Boston University) 178(1) (1996): 13–34.

Schifter, Deborah. "Reasoning about Operations: Early Algebraic Thinking, Grades K through 6." In *Mathematical Reasoning, K–12*, 1999 Yearbook of the National Council of Teachers of Mathematics (NCTM), edited by Lee V. Stiff and Frances R. Curio, pp. 62–81. Reston, VA: NCTM, 1999.

Yerushalmy, Michal. "Making Exploration Visible: On Software Design and School Algebra Curriculum." *International Journal of Computers for Mathematical Learning* 4(2–3) (1999): 169–89.

Activity 1

Given the points $A(-3, 2)$ and $B(5, 1)$
  (a) Find the length AB.
  (b) Write an equation of the perpendicular bisector of AB.
  (c) Write an equation of the set of points (x, y) such that the sum of the distances from (x, y) to A and B is 9.

Solution:

- $\sqrt{(a2-a1)^2 + (b2-b1)^2} \rightarrow \text{dist}(a1,b1,a2,b2)$ Done

- dist(¯3,2,5,1) $\sqrt{65}$

- solve(dist(x,y,¯3,2) = dist(x,y,5,1), y) $y = \dfrac{16x-13}{2}$

- solve(dist(x,y,¯3,2) + dist(x,y,5,1) = 9,y)

$$y = \dfrac{\left(9\sqrt{x^2 + 2x + 19} + 2(x-16)\right)}{20}$$

or

$$y = \dfrac{-\left(9\sqrt{x^2 + 2x + 19} - 2(x-16)\right)}{20}$$

Line 1: Uses the distance formula to define a function that gives the distance between two points (a_1, b_1) and (a_2, b_2).

Line 2: Uses the formula to find the length of the segment.

Answer: $\sqrt{65}$

Line 3: The perpendicular bisector of a segment is the set of all points equidistant from the endpoints of the segment. Setting the distances of the endpoints from (x, y) and solving for y gives the equation of the perpendicular bisector.

Answer: $y = \dfrac{16x-13}{2}$

Line 4: Gives the equations of the two halves of the ellipse in a form that could be used for graphing.

Note: The problem above tests: (1) students' knowledge of the distance formula by using it in a basic way (see **part a**), (2) students' knowledge of the definition of perpendicular bisector and the ability to apply it by using the distance formula (see **part b**), and (3) students' ability to use the distance formula in a different context (see **part c**). Nowhere does the problem stem mention what formula to use or how to use it. Furthermore, the stem does not define *perpendicular bisector of a segment*, so the student must know this definition. The directions in part (c) do not indicate what type of curve to expect—an oblique ellipse given by a two-equation definition. Consider the difficulty of the algebra involved, especially in part (c).

The Point: Students need to know a great amount of math to do this problem; they do not need to perform any symbol manipulation.

CHAPTER 2

Theories for Thinking about the Use of CAS in Teaching and Learning Mathematics

M. Kathleen Heid

TWENTY YEARS ago, I first had the opportunity to experiment with a symbolic manipulation program in mathematics instruction. Unlike computer algebra systems that, within a single program, generate graphs, tables, and curves of best fit, the muMath software program that I first used in my introductory calculus class was a symbolic manipulation program that processed symbolic expressions and operated on an Apple II Plus (48K) computer with a CPM card. The students enrolled in my calculus class used muMath in conjunction with separate computer programs that I had written for graphing, curve fitting, and production of tables of function values. Prior to class each day, I rolled two carts into the classroom; each cart held an Apple II Plus computer that I used for demonstration. Outside of class, the approximately forty students shared four Apple II Plus computers that were stored in a metal closet in a mathematics professor's office.

At that time, I described what I was doing in a somewhat primitive way: my students were able to use the computer as a tool to learn mathematics. I hypothesized that they could better direct their attention to the formulation and interpretation aspects of mathematics if they "outsourced" the computational tasks to the computer. Therefore, I directed my students' attention to the concepts of calculus and taught them to use the computer—to generate tables and graphs; to evaluate limits, derivatives, and integrals; to find exact solutions for equations; and to find curves of best fit. Peschek and Schneider (2001) refer to the instructional approach described above as "an interplay between representing, operating, and interpreting" (p. 8).

Today—twenty years later—I expect each of my undergraduate students, who are prospective secondary mathematics teachers, to have personal access to a TI-92 Plus calculator. The TI-92 Plus calculator has more than 180K of user-available RAM, an additional 380K of Flash ROM, menu-driven access, and many more capabilities than the symbolic manipulation program I first used in my classroom. I still ask my students to explore families of functions using the graphing, table generation, and symbolic manipulation capabilities of the CAS. I supplement CAS technology with new representations available through dynamic geometry tools and hope to have access to similar statistical tools in the near future. My colleagues who teach in high school classrooms also use CAS to achieve similar learning goals. One of the most significant changes in the past twenty years, aside from the astounding increase in the power of the computer/calculator, has been the development and refinement of theories related to the use of tools, such as CAS, in teaching and learning mathematics.

I write this chapter, not as an authority on theories of learning but as an educator drawn to a range of theories that seem promising in explaining the impact of the use of CAS in mathematics education. I focus on theories that "inform" my understanding of learning as it relates to the use of CAS and share some ideas that may help mathematics teachers think about teaching and learning mathematics using CAS.

How Theories Related to Mathematics Learning and Teaching Can Inform Us in Using CAS

How is the process of learning mathematics with CAS different from the process of learning mathematics in other technological environments? What do the theories of mathematics teaching and learning suggest about the role of CAS in mathematics education? The CAS is more than simply another computer or calculator program, and access to a CAS does more than offer students an alternative strategy for problem solving. CAS have the potential for being what Pea (1987) has described as a "cognitive technology," a medium that helps "transcend the limitations of the mind ... in thinking, learning, and problem-solving activities" (p. 91).

I can think of at least three ways in which the CAS can function as a cognitive technology: (1) Students can use CAS for the repeated execution of routine symbolic procedures in rapid succession, without the diminished accuracy and increased fatigue usually associated with the repetitive execution of by-hand algebra routines. Although this capacity for repeated execution of routine procedures is the hallmark of a wider range of tools, including spreadsheets and graphing utilities, using CAS for repeated routine symbolic procedures replaces the symbolic activity usually performed by high school mathematics students. Therefore, students can depend on CAS production if they want to search for patterns in symbolic results. For example, students may repeatedly expand polynomials of the form $(a + b)^n$ for different values of n in their search for a pattern in the coefficients or they may expand $(a + b + c)^n$ for different values of n to generate data for a search for a symbolic pattern in the terms of successive expressions. (2) Students can assign rote symbolic tasks to the CAS so that they can concentrate on making "executive" decisions. For example, when solving a complicated problem, students can allocate the equation solving or the exact evaluation of complicated expressions to the CAS so that they can spend more time crafting an approach to solving the problem. (3) Students can use the CAS to apply routine symbolic algorithms to complicated algebraic expressions, without the confusion students sometimes experience when trying to apply a routine procedure to a complicated expression. Once a function is defined, students use the same keystrokes to perform a procedure on an equation involving a simple function as to perform the procedure on an equation involving a more complicated function (see **fig. 2.1**). Students can direct their attention to more global algorithms, strategies, concepts, and applications rather than to the details of executing symbolic routines; they can benefit from an increased—or "ramped up"—level of abstraction without losing touch with concrete examples. CAS can assume the role of a cognitive technology, either in single instances or in a short series of instances, but CAS have the greatest impact on student learning when they assume a central role in the mathematics curriculum.

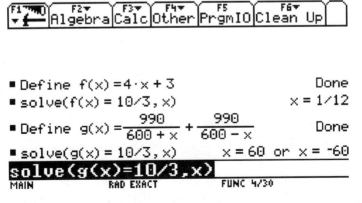

Fig. 2.1. A procedure applied to two functions that differ in complexity

Uses of CAS That Influence the Content, Priorities, and Sequence of Curricula

The types of theories that have been particularly useful to me in clarifying the potential roles of CAS in mathematics learning include the following: (1) theories that relate the *structure* of the mathematics curriculum to student learning, and (2) theories that relate the *content* of the mathematics curriculum to student learning.

Using CAS in instruction can influence the structure and the content of the mathematics curriculum. I define "structure" to characterize the way in which a curriculum is sequenced and the types of activities that are emphasized in the curriculum. A few types of theories that are useful when considering the structure of school mathematics are as follows: theories that relate to the relationship between procedures and concepts, theories that relate to the role of superprocedures and subprocedures, and theories that relate to using scaffolding. The new perspectives and theories related to mathematics content that I have found to be useful are those that focus on the components of school algebra and introductory calculus. Theories that look at new ways for students to encounter and develop an understanding of different mathematical concepts (e.g., function, variable, equivalence, system, limit, derivative, integral) are also useful. Each of the different perspectives on school algebra and on introductory calculus carries with it different potential roles for the CAS and calls for theories addressing the learning of different types of mathematical subject matter. Such perspectives include theories about the transition from informal to formal mathematics, as well as theories about the progression from process to object understandings.

Theories Related to the Structure of the Mathematics Curriculum

With the help of current CAS technology, students are able to access high-level mathematical processes previously inaccessible to them. As students engage in problem solving, they can use CAS to generate and manipulate symbolic expressions that were otherwise too time-consuming, too tedious, or too complicated. Although CAS can be used to address a variety of instructional goals, their actual impact on teaching and learning mathematics depends on the role of CAS in specific mathematics instruction.

CAS as Amplifiers and Reorganizers

Cognitive technologies, such as CAS, assume a variety of distinct roles in mathematics instruction. Pea (1985, 1987) suggests two ways that CAS and other technologies can function. He states that such technologies can function either as amplifiers or as a reorganizers and can afford the user the opportunity to "transcend the limitations of the mind" by their ability to generate more examples and a greater range of examples for students. The CAS may be used to facilitate students' learning of the mathematics content they would have studied without using the CAS, while leaving the goals and sequence of the curriculum basically intact. Pea (1985, 1987) defines an amplifier as a technology used to extend the existing curriculum. Instructors who use CAS as amplifiers may give students access to CAS calculators in class but not on examinations, thus expecting them to show mastery of the by-hand procedures that they would have learned without using the CAS. For example, instructors may allow students to use the CAS to check answers or to perform algorithms only after they have shown that they can produce the same answers with by-hand procedures. Instructors may also encourage students to use the CAS to generate data from which they could infer procedural rules that are new to them. The existing curriculum would not change in content, but the amplified curriculum would help students accomplish the current objectives more quickly, thus *extending* the curriculum.

CAS can also function as cognitive technologies when serving as reorganizers, thus changing the fundamental nature and arrangement of the curriculum. Kilpatrick and Davis (1993, p. 204) emphasized the importance of technology in its role as a reorganizer:

> The computer is not merely an amplifier of general curriculum issues specific to the mathematics curriculum. ... It changes certain fundamental questions one needs to consider in any attempt to revise the mathematics curriculum by making the subject matter itself more problematic. What is mathematics? What knowledge of mathematics does tomorrow's society demand? What mathematics should this pupil learn so as to be a wise and humane citizen of that society?

Many mathematics educators have made the unquestioned assumption, at some point in their careers, that students must first master the procedures of mathematics before they can handle the conceptual ideas of mathematics. This assumption often manifests itself in the form of an objection by the teacher to using technology in the classroom, such as the following: "I don't let my students use calculators to do X. How can they really understand X if they don't know what the calculator is doing?" The objecting teacher does not mean to imply that students should become facile with the specific algorithm used by the calculator, an accomplishment that may be inaccessible to both teacher and students, but intends to state that students should be able to perform the given task without a calculator. The assumption that procedural fluency must precede concept acquisition is a theory, not a fact, and, as such, can be tested. An opposing theory is actually more widely supported in research studies using the CAS. The theories that I find to be most useful are those that have been tested through research studies.

Challenging the Assumption of Procedures before Concepts

Because CAS can be used in the role of reorganizer to facilitate adjustment in the balance, sequence, and priorities assigned to concepts and procedures in the mathematics curriculum, they have frequently been used to test the theory that skills development must precede concept development in mathematics instruction. In some studies that focus on the CAS as a reorganizer, students in calculus classes (Heid 1988; Palmiter 1991) and in introductory algebra classes (Heid 1992; Heid et al. 1988; Matras 1988; Oosterum 1990; Sheets 1993) used the CAS for almost all of the routine symbolic manipulations, for production of graphs and tables of values, and for curve fitting. In these studies the nature of the calculus curriculum and the nature of the algebra curriculum were fundamentally changed. Student time traditionally spent performing routine symbolic manipulation procedures was reallocated to interpreting symbolic results and to recognizing and applying algebra and calculus concepts. In the algebra classes, for example, students studied the properties and applications of prototypical families of linear, quadratic, exponential, and rational functions instead of spending time on by-hand factoring and simplifying rational expressions. In the previously mentioned studies, concepts and applications were no longer what one did before or after the "real mathematics" of solving equations or rewriting algebraic expressions. Concepts and application became the *centerpiece* of mathematical activity. The research conducted in these settings examined students' understandings of mathematical concepts—such as variable, function, derivative, and integral—that are specific to algebra and calculus; it also examined the students' abilities to reason about mathematical models and to solve algebraic problems.

Studies, such as those mentioned in the previous paragraph, that cast the CAS in the role of reorganizer have been used to test the theory that the ability to perform procedures must precede the development of conceptual understanding. Have the technological procedures in the "reformulated curricula" mentioned above simply replaced by-hand procedures so that the result is merely a substitution of skills rather than a counterexample to the skills-before-concepts assumption? Even if "yes" is the answer to the preceding question, this substitution of skills seems to allow students more time to gain a better understanding of other mathematics. In the instance of studies cited in the preceding paragraph, using a CAS before developing related by-hand skills seemed to help students learn concepts in greater depth than the traditional skills-before-concepts curricula. We need to be careful to avoid interpreting the results of these studies to infer that *any or all* use of the CAS results in better conceptual understanding for students. It may be more appropriate to conclude that CAS allow for a restructured curriculum that fosters students' understanding of concepts. Student success may be attributed to the additional time dedicated to the development of conceptual understanding and to the focus on concepts facilitated by outsourcing manipulative skills.

The role of technology as a reorganizer has been investigated in a range of classroom studies related to mathematics learning. However, instructional uses of CAS cannot be easily categorized into exactly one of these classifications, nor can the instructional uses that are labeled as amplifiers or reorganizers be fully described by these categories. The concepts of amplifier and reorganizer are useful, however, in highlighting important differences among classroom environments in which CAS has assumed a central role. As we reflect on the variety of CAS uses developed by educators, we should notice whether the CAS is used to enhance the delivery of a curriculum whose major objectives remain unchanged or whether the CAS is used to substantially change the content of the curriculum.

The CAS, Macroprocedures, and Microprocedures

Another theory that is useful as we reflect on the role of the CAS in mathematics instruction is drawn from a cognitive science perspective. Davis (1984) described the causes of students' errors as mathematical superprocedures and subprocedures. He pointed out that students' errors often come not from errors in executing subprocedures but from the incorrect choice of a subprocedure. Davis' concept of mathematical superprocedures and subprocedures can be viewed in terms of what I have called macroprocedures and microprocedures. Mathematical procedures can be viewed as sequences of macroprocedures whose major procedural steps are the component microprocedures. For example, the process of solving a quadratic equation in x can be thought of as a macroprocedure. Its three component steps—or microprocedures—are as follows: (1) reexpressing the equation so that its right member is zero, (2) expressing the left member as a product, and (3) finding the zeroes of each of the factors of this product. When CAS are available, a variety of combinations of menu-driven microprocedures are possible for a single macroprocedure. For example, the macroprocedure for finding the zeroes of quadratic functions might consist of the two following microprocedures: (1) identifying the values of the parameters, a, b, and c, and (2) executing a "solve" command, such as SOLVE ($ax^2 + bx + c = 0, x$). Other sets of microprocedures could also correspond to the aforementioned macroprocedure. The CAS has, for the most part, menu commands for executing microprocedures at a variety of different levels (e.g., commanding a step-by-step solution of equations versus using a "solve" command).

Using the CAS in the context of macroprocedures and microprocedures is an example of "scaffolding," as described by Kutzler. Kutzler (2000) describes scaffolding in general terms: "The *scaffolding method* is any pedagogically justified sequence of using and not using technology for trivialisation, experimentation, visualisation, or concentration either in the sense of automation or compensation" (p. 18). A mathematics curriculum that gives students the responsibility for generating macroprocedures and allows them access to the CAS for executing microprocedures is using a scaffolding method. Students who understand the macroprocedures seem to have a framework within which to place the microprocedures when and if they learn them. The students who were enrolled in the algebra and calculus classes described previously in this section learned the macroprocedures for solving problems long before they had mastered the component microprocedures. For example, the algebra students solved quadratic equations before they learned to factor, and the calculus students maximized functions of one variable before they learned to take derivatives by hand. When, at a later date, these same students learned procedures for factoring and derivatives, their exposure to scaffolding provided them with something on which to "hang the skills" they were developing. Consequently the students' mathematical development seemed to proceed more rapidly.

Theories Related to the Content and Processes of the Mathematics Curriculum

The theories of learning and knowing that I have discussed in the preceding sections influence the ways in which the CAS can open the door to significant reordering of content and to shifts in emphasis in school mathematics. However, these theories do not significantly alter the nature of the mathematical content. A different, but not entirely unrelated, collection of theories and perspectives addresses the impact of the CAS on the content of the mathematics curriculum in a more fundamental way. First, some theories address the impact of the CAS on the learning of school algebra in its various incarnations. Second, other theories address the impact of the CAS in curricula as students move from using informal to formal strategies. Third, still other theories address the impact of CAS on the level of the mathematical object that students study.

CAS and School Algebra

Algebra is the area of school mathematics that has the most potential to be affected by CAS, because CAS perform most school algebra routines quickly and on command. This potential raises the question "What algebra should be taught in schools?" The answer to the preceding question depends on one's definition of "school algebra."

Bednarz, Kieran, and Lee (1996, p. 4) outline four approaches to school algebra:

> These options, which are at the forefront of contemporary research and curriculum efforts on the introduction of secondary school algebra, are as follows: generalizing patterns (numeric and geometric) and the laws governing numerical relations ..., solving of specific problems or classes of problems ..., modeling of physical phenomena ..., and focusing on the concepts of variable and function.

A natural addition to these four approaches is a perspective that focuses on the structure of algebra as it engages students in reasoning about abstract entities governed by a fixed set of properties. Each of the now five perspectives places importance on a different aspect of algebra, and different theories of knowing and learning seem to be generally compatible with each of the five different perspectives. Theories that address the learning of procedures and abstractions are more likely to shape the view of school algebra as the study of *structure*. Theories that address the learning of concepts may be more likely to shape the view of school algebra as the study of *variable* and *function*.

I have found it interesting to examine the impact of particular perspectives on algebra as they relate to the role of the CAS in mathematics instruction. For example, consider the perspective of algebra as generalization. CAS offer new access to the *generalization* of symbolic results, but the ways in which a curriculum takes advantage of the capacity of CAS to generalize symbolic results depend on the theories of learning and knowing that are brought into play. The belief that learning occurs through students' deliberate action and their reflection on that action suggests that students should generate instances of a pattern and also reflect on the meaning of those results. For example, a teacher who believes in the importance of deliberate action and reflection would require that students who use CAS move beyond the observation that for each numerical value of k, the constant term in the expansion of $(a + 1)(a + 2)(a + 3) \ldots (a + k)$ has a value of $k!$ (see **fig. 2.2**). Similarly, a teacher would require those students who notice that the second term of the expansion is the triangular number

$$\sum_{1}^{k} i$$

to reflect on *why* that pattern holds. As a general rule, evidence of appropriate reflection would include the requirement that students generate a supported conjecture to explain *why* a pattern that they "discover" proves to hold true.

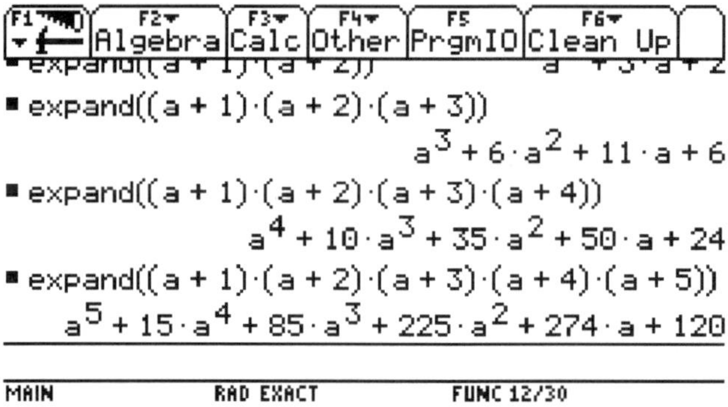

```
F1    F2      F3    F4    F5       F6
  Algebra Calc Other PrgmIO Clean Up
```
- expand((a + 1)·(a + 2)·(a + 3))
  $$a^3 + 6 \cdot a^2 + 11 \cdot a + 6$$
- expand((a + 1)·(a + 2)·(a + 3)·(a + 4))
  $$a^4 + 10 \cdot a^3 + 35 \cdot a^2 + 50 \cdot a + 24$$
- expand((a + 1)·(a + 2)·(a + 3)·(a + 4)·(a + 5))
  $$a^5 + 15 \cdot a^4 + 85 \cdot a^3 + 225 \cdot a^2 + 274 \cdot a + 120$$

MAIN   RAD EXACT   FUNC 12/30

Fig. 2.2. Instances of the expansion $(a + 1)(a + 2)(a + 3) \dots (a + k)$ on a TI-92 Plus

Particular theories of learning and knowing can coexist with more than one different perspective on algebra, and each viable combination of learning theory and algebra perspective may play out in a different way in the algebra classroom. As an educator, I might believe a theory of learning that held the view that students' primary source of learning was through direct experience with mathematical representations and through reflection on that experience. Simultaneously, I might have a functions perspective on algebra. In this context, if my goal were to help students develop an understanding of parameters, I might design activities that would engage students in investigating the effects of parameters on different families of functions. I would design the explorations so that students would have the opportunity to generate and test their own conjectures about the effects of different parameters. The students would learn by reflecting on the results of their actions, and CAS would foster the students' reflection on the role of the symbolic in their observations. Some of the curricula that have been created, such as *Concepts in Algebra: A Technological Approach* (CIA) (Fey et al. 1995, 1999) and *CAS-Intensive Mathematics* (CAS-IM)—a curriculum project currently under way and codirected by Rose Zbiek and M. Kathleen Heid—are designed to engage students during their early high school years in explorations similar to those previously described. In designing the previously mentioned CIA and CAS-IM curricula, we assumed that students develop understandings of the "big ideas" of mathematics through continued and varied experiences with the ideas and reflections on their experiences.

Two individuals with similar perspectives on algebra will not necessarily hold the same beliefs about learning. I recall a conversation with a parent whose daughter had encountered the graph of a quadratic function at the beginning of an introductory algebra course that used the CIA curriculum. He was pleased with the central focus on functions and on his daughter's ability to use functions to describe real-world settings, but he had some concerns about how we, the curriculum developers, were approaching the topic. In the second chapter of the CIA text, his daughter had encountered a parabolic graph—a topic that would not be studied in any serious depth until the fourth chapter in the CIA textbook. Compensating for what he believed was a deficit in the textbook, he explained to his daughter how to produce a graph of a function of the form $f(x) = ax^2 + bx + c$. He informed her that the graph of every function in the form $f(x) = ax^2 + bx + c$ is shaped like a parabola, that the parabola will curve upward if a is a positive number, and that the parabola will cross the vertical axis at $(0, c)$. He valued the function approach and offered me his advice on ways to better structure the curriculum, but his advice ran contrary to my view of learning.

The developers of the CIA curriculum intended for students to have a variety of experiences with graphic, numeric, and symbolic representations of curves of various shapes early in the course and that they would learn about each family of graphs as they acquired more experience. The interested parent described above believed that each concept should be mastered as it was encountered. Furthermore, he believed that students who are introduced to quadratic functions should not leave the topic until they have achieved an acceptable level of skill with the topic. The parent thought that we, as the curriculum developers, should show students how to graph the functions by hand and allow them to generate graphs with technology only after they had mastered the by-hand technique. The differences between our theories of learning and those of the parent suggest vastly different approaches to using the CAS in the service of a functions perspective on algebra.

CAS and Students' Movement from Informal to Formal Strategies

As a teacher of high school mathematics and as a mathematics education researcher and curriculum developer, I have searched continually for ways to help students move from their informal knowledge of algebra to a more formalized knowledge of algebra. Theories of learning and knowing take into consideration the difficulty that students have in developing an ability to operate in a closed formal mathematical system. Will CAS facilitate or hinder this development? The question remains unanswered, but the answer is of particular concern in the context of a computer algebra environment. CAS require students to enter and read formal symbolic language, with little or no tolerance for deviations nor acceptance of user-designed notations. Computer algebra systems, with or without graphical and numerical capabilities, require the user to communicate in a language specific to the CAS and different from paper-and-pencil language. This need is perhaps most salient in the symbolic manipulation portion of the CAS, as manifested in its unyielding demand for perfectly syntaxed commands. The CAS will use any legitimate input provided by the user but does so on pre-established expectations. A user who intends to enter and use the expression

$$\frac{x^2 - x - 6}{x - 3}$$

but enters $(x^2 - x - 6) / x - 3$ will, perhaps unknowingly, be operating instead on

$$\frac{x^2 - x - 6}{x} - 3.$$

In their current forms, even menu-driven symbolic manipulation programs require some knowledge of acceptable inputs for each command. For example, equation-solving commands require the user, at minimum, to enter the equation and the variable for which the equation is to be solved. The limit commands often require input of the symbolic function rule or expression—as well as the limiting value of the variable and the direction from which the limit is to be approached. Although CAS programs may provide templates for the form of the input, a student who uses the templates may first be required to understand the meaning of the component mathematical entities (e.g., to calculate

$$\lim_{x \to 3^-} \left(\frac{x^2 - x - 6}{x - 3} \right),$$

the TI 89/92 Plus user must understand that the template "LIMIT(EXPR, VAR, POINT [,DIRECTION])" requires the entry of an expression like "limit((x^2 − x − 6)/(x − 3), x, 3, −1))." How can students who are operating at the informal stages of algebra cope with such a substantial demand in using a CAS to communicate in a strictly formal language? In a situation such as the one just described, theories of learning and knowing do not provide immediate answers; instead they highlight the importance of, and difficulties associated with, students' movement toward work with formal symbolic systems.

The requirement of the CAS for exact language input calls to mind a distinction that researchers (Doerr 1995; Gravemeijer et al. 2000) have made between "expressive" and "exploratory" approaches to computer-based instructional design. Exploratory approaches are defined as approaches in which "students explore conventional mathematical symbolizations in experientially real settings" (Gravemeijer et al. 2000, p. 228) with the instructional purpose of "students' development of the mathematical understandings inherent in the mature use of symbolizations" (Gravemeijer et al. 2000, p. 228). Expressive approaches, by contrast, are based on the assumption that students will invent "increasingly sophisticated symbolizations with limited guidance" (Gravemeijer et al. 2000, p. 228). The two approaches have different starting points. The exploratory approach starts with conventional notation, and the expressive approach starts with the student's informal notation. When students are working with paper and pencil, they are presumably more "in control" of the representations they use than when they are working with CAS. In other words, pencil-and-paper notation can behave the way students want it to behave. In contrast, students using CAS need to communicate in the language of the CAS and may need to forego their own informal notation or language. The CAS environment lends itself much more to an exploratory approach than to an expressive approach. The distinction between the exploratory and expressive approaches presents a useful way to think about instruction that integrates the CAS and raises issues regarding the effects of CAS in mathematics instruction. What effect will the need to operate in the formal world of the CAS have on the extent to which students build their own understandings and personally meaningful informal strategies? What effect will the process of involving students in the notational culture of the mathematician have on students' abilities to interpret and reason from this notation? What initial or additional work with symbols and other representations may be needed to develop students' ownership of the symbols they use? As more students use CAS in the development of their mathematical ideas, answers to these questions are likely to come into greater focus.

CAS-intensive Curricula That Can Assist Students As They Move from Informal to Formal Strategies

As I consider the shape of curriculum influenced by the CAS, one theory of learning and instruction that comes to mind is that of Realistic Mathematics Education (RME). RME is based (Meyer 2001) in a view of learning referred to as a constructive activity—one that relies on students' participating in, and reflecting on, experientially real activities in sociocultural settings (Gravemeijer et al. 2000). RME philosophy contends that learning moves gradually through different levels of abstraction as it uses models, diagrams, tables, and symbols and proposes that mathematical understanding is structured and connected. Drijvers and van Herwaarden (2000) articulate a question about the capacity for the CAS to support RME instruction: "The question is ... whether computer algebra supports gradual formalisation of informal strategies" (p. 259).

The role of CAS in an RME curriculum is problematic. Given the importance of students' developing their own representations, we might naturally ask whether the CAS could possibly have a role in a curriculum that is not designed in the context of RME but that, instead, takes seriously the importance of building on students' informal understandings. A curriculum that assumes students' access to CAS can conceivably be crafted to help students make the transition from informal to formal settings. For example, one can use, as a starting point, the CIA curriculum (Fey et al. 1995, 1999), a curriculum that was premised on the availability to students of a technology toolkit with the capability of a computer algebra system. As curriculum developers, our approach to the concept of equivalence, a central concept in the CIA curriculum, is one example of the possible role that CAS can play in the development of formal algebra. Theories that rely on the development of hypothetical learning trajectories have begun to play important parts in research studies focused on understanding students' mathematical understandings. Hypotheses about ways in which students' understandings of equivalence may develop were central to this part of the CIA curriculum design and delivery. First, students' understanding of equivalence of symbolic expressions may start with their observation that two seemingly different rules describe what they know to be the same phenomenon. This understanding might evolve into reasoning about the structures in context but without the necessity of referring to the situation. Finally, students' understanding could take the shape of reasoning about symbolic expressions that are not connected with a real world context.

The CIA curriculum was designed to ensure that students acquire extensive experience with graphic, numeric, and symbolic representations of functions prior to considering equivalence from a formal or symbolic perspective. By the time students had reached the final two chapters of the book, "Chapter 8: Symbolic Reasoning: Equivalent Expressions" and "Chapter 9: Symbolic Reasoning: Equations and Inequalities," they had engaged regularly in constructing and using graphic, numeric, and symbolic representations of functions. Until that point, they had used symbolic representations to answer questions about the context but had been required to do little reasoning from the symbolic representations. Chapter 8 built on those activities with a situation that required them to symbolize and analyze such situations as the Pet Wards scenario adapted in figure 2.3.

> **Situation 1.1.** A national pet-hotel chain is planning to build units for use in a series of franchises. Each of the units for small pets is two rows of two-meter-by-two-meter square wards. The wards are connected as shown in the partial floor plan below.

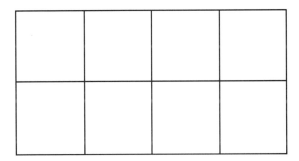

> Walls for these units come only in two-meter panels, and the number of two-meter panels needed depends on the number of wards to be included in the unit. Because the management plans to build many units of different sizes, the manager wants to have a rule relating the number of wards and the number of panels.

Fig. 2.3. Activity from the CIA curriculum that requires students to symbolize a contextual situation and defend their symbolic representation (Fey et al. 1995, 1999, p. 391)

This context of the Pet Wards situation was used as a site for students to generate and determine the equivalence of a variety of potential function rules. Since students had developed a comfort level with respect to contextually situated function rules, they were expected to soon be ready to reason about forms without needing the support of contextual situations. The movement toward students' not requiring contextual support is one of the premises of the RME curriculum work. The first seven chapters of the CIA textbook treated symbolic representations in the background, primarily to represent and to generate numerical and graphical representations. The last two chapters of the CIA textbook brought conventional symbolic representations to the foreground. Before this point in the CIA textbook, students were required to do very little by-hand symbolic manipulation. While working on the Pet Wards situation in which n represents the total number of wards needed, students might argue that is actually the same as

$$5\left(\frac{n}{2}\right)+2$$

$$7+5\left(\frac{n}{2}-1\right)$$

because

$$5\left(\frac{n}{2}\right)+2$$

represents juxtaposing a string of $\frac{n}{2}$ sideways-E-shaped modules and finishing off the last one with a double vertical panel. Similarly,

$$7+5\left(\frac{n}{2}-1\right)$$

represents the same activity in a slightly different order, starting with a seven-sided double ward and juxtaposing a string of $\frac{n}{2}-1$ Ǝ-shaped modules. The students can use CAS as computational aids as they reason about complicated algebraic structures and gather evidence about the equivalence of expressions through graphs and tables, verifying through "simplifying" symbolic expressions and considering the reasons two expressions may or may not be equivalent on the basis of properties of real numbers. Thus, a curriculum that starts from the premise of context-based problem solving for students can, with the assistance of CAS, evolve into one that has students study and learn from symbolic work that is not contextually driven.

CAS and the Development of Process and Object Understandings

Theories regarding the nature of understanding of specific mathematical concepts—including the distinction and interaction between process and object understandings of a mathematical concept—are central to the work of explaining mathematical thinking at the secondary and postsecondary level. Students who possess a "process understanding" of a mathematical concept think of the concept only in terms of a procedure. For example, students with a process understanding of function might think of a function only in terms of a rule for obtaining output values. Students with a process understanding of function would recognize $f(x) = 8x + 9$ in terms of multiplying an input value by 8 and adding 9, and would view

$$g(x) = \frac{x}{2}$$

in terms of dividing the input value by 2 to obtain the output. Those students, who would not be attending to the linear nature of either function, would not be inclined to think of the sum, $f + g$, as a linear function.

In contrast, students with an object understanding of function can operate on functions as entities; they can compose two functions and think of the result as another function. The process-object debate is so significant to the subject of mathematical understanding that it frequently occupies center stage in discussions related to this topic. Mathematics education researchers are reluctant to view process and object as separable and have focused their attention on characterizing the nature of the relationship between these understandings (Sfard 1991; Gray and Tall 1994). As these researchers have pointed out in various ways, a seemingly more reasonable view sees concepts and procedures as influencing each other. One can argue that successful students move back and forth between their process and object understandings as they work on mathematics problems.

Technology has played an important role in research studies that characterize students' process and object understandings of mathematical concepts. The combination of a computer algebra system and ISETL—a mathematical programming language that accepts function rules as input and gives function rules as output—has been used in studies to foster acquisition of an object understanding of function and a range of other concepts from postcalculus mathematics. Dubinsky and his colleagues (Asiala et al. 1991) have developed the APOS (Action-Process-Object-Schema) theory to describe the progression from action to process to object levels of understanding. At the action level students perform a procedure step by step, often cued by the results of one step to execute the next. At this level, students are simply "going through the motions" and are unable to describe the procedure without actually performing it. At the process level, students can describe the steps of the procedure without actually executing the procedure. Students working at the object level can apply actions to processes. For example, if students have an object-level understanding of function, they can apply the action of forming a composition to two functions and understand that the result is another function on which they can operate.

CAS are ideal technologies for fostering growth in the level of understanding of mathematical concepts. Using the CAS, students can signify and execute actions on mathematical entities without carrying out the procedures by hand. A computer algebra system that allows the user to craft original function rules and small programs acts as a tool for capturing processes so that actions can be performed on them. A calculus student, for example, can use CAS to optimize a function of one variable by defining a function f of one variable and then executing the commands

$$\frac{d}{dx} f(x)$$

and

$$\text{Solve}(f(x) = 0, x)$$

and evaluating f at the solution values—perhaps at values in the neighborhood of the solution. The sequence of commands is the same no matter what f is being considered. Students may be better able to internalize the process of optimization, developing the process into an object, without being distracted by the obligation of simultaneous by-hand symbolic manipulation. I do not claim that a student who can write a routine to perform a procedure can operate on that procedure as an object. Rather, I suggest that the availability of this accurate and convenient tool for condensing a process into a single command has the potential to serve as an intermediate step between process and object understandings. If a student can write a routine as a single function, the next step might be to think of the process represented by the function as an entity that can be manipulated. However, an action performed on a physically available symbol must be distinguished from and an action performed on a mathematical entity that the student has constructed mentally.

Conclusion

Theories related to learning of mathematics can help us as educators understand the ways in which CAS fundamentally change the environment in which mathematics is learned, along with the possible impact of those changes. The computer algebra system, with its symbolic manipulation capacity, provides both an aid and a challenge to students as they develop solid understandings of mathematical concepts and processes. By allowing students to manipulate such mathematical entities as variable expressions, function rules, and equations, the computer algebra system gives them access to the tools of mathematics. *Therein lies the assistance*. But we should remember that the gift of true "power" obtained through using the tools of mathematics demands a deep-level understanding of the objects on which the tools operate. The effectiveness of any tool depends entirely on the decisions made regarding its use. *Therein lies the challenge!*

Note: An earlier version of this paper was presented at the CAME 2001 Symposium: *Communicating Mathematics Through Computer Algebra Systems* in Utrecht, The Netherlands.

REFERENCES

Asiala, Mark, Anne Brown, David J. DeVries, Ed Dubinsky, David Mathews, and Karen Thomas. "A Framework for Research and Curriculum Development in Undergraduate Mathematics Education." In *Research in Collegiate Mathematics Education*, vol. 2, edited by James Kaput, Alan H. Schoenfeld, and Ed Dubinsky, pp.1–32. Providence, R.I.: American Mathematical Society, 1991.

Bednarz, Nadine, Carolyn Kieran, and Lesley Lee. *Approaches to Algebra: Perspectives for Research and Teaching*. Dordrecht, The Netherlands: Kluwer Academic Publishers, 1996.

Davis, Robert B. *Learning Mathematics: The Cognitive Science Approach to Mathematics Education*. Norwood, N.J.: Ablex Publishing Corporation, 1984.

Doerr, Helen M. "An Integrated Approach to Mathematical Modeling: A Classroom Study." Paper presented at the annual meeting of the American Educational Research Association, San Francisco, April 1995.

Drijvers, Paul, and Onno van Herwaarden. "Instrumentation of ICT-Tools: The Case of Algebra in a Computer Algebra Environment." *International Journal of Computer Algebra in Mathematics Education* 7(4) (2000): 255–75.

Fey, James T., M. Kathleen Heid, Richard A. Good, Glendon W. Blume, Charlene Sheets, and Rose Mary Zbiek. *Concepts in Algebra: A Technological Approach*. Chicago: Everyday Learning Corporation, 1995, 1999.

Gravemeijer, Koeno, Paul Cobb, Janet Bowers, and Joy Whitenack. "Symbolizing, Modeling, and Instructional Design." In *Symbolizing and Communicating in Mathematics Classrooms: Perspectives on Discourse, Tools, and Instructional Design*, edited by Paul Cobb, Erna Yackel, and Kay McClain, pp. 225–73. Mahwah, N.J.: Lawrence Erlbaum Associates, 2000.

Gray, Eddie, and David O. Tall. "Duality, Ambiguity, and Flexibility: A Proceptual View of Simple Arithmetic." *Journal for Research in Mathematics Education* 25 (March 1994): 116–40.

Heid, M. Kathleen. "Resequencing Skills and Concepts in Applied Calculus Using the Computer as a Tool." *Journal for Research in Mathematics Education* 19 (January 1988): 3–25.

———. *Final Report: Computer-Intensive Curriculum for Secondary School Algebra*. Final report for NSF project number MDR 8751499. University Park, Pa.: Pennsylvania State University, Department of Curriculum and Instruction, 1992.

Heid, M. Kathleen, Charlene Sheets, Mary Ann Matras, and James Menasian. "Classroom and Computer Lab Interaction in a Computer-Intensive Algebra Curriculum." Paper presented at the annual meeting of the American Educational Research Association, New Orleans, La., April 1988.

Kilpatrick, Jeremy, and Robert B. Davis. "Computers and Curriculum Change in Mathematics." In *Learning from Computers: Mathematics Education and Technology*, edited by Christine Keitel and Kenneth Ruthven, pp. 203–21. Berlin: Springer-Verlag, 1993.

Kutzler, Bernard. "The Algebraic Calculator as a Pedagogical Tool for Teaching Mathematics." *International Journal of Computer Algebra in Mathematics Education* 7(1) (2000): 5–23.

Matras, Mary Ann. "The Effects of Curricula on Students' Ability to Analyze and Solve Problems in Algebra." Doctoral dissertation, University of Maryland, 1988.

Meyer, Margaret R. "Representation in Realistic Mathematics Education." In *The Roles of Representation in School Mathematics*, 2001 Yearbook of the National Council of Teachers of Mathematics (NCTM), edited by Albert A. Cuoco and Frances R. Curcio, pp. 238–50. Reston, Va.: NCTM, 2001.

Oosterum, M. A. M. Boers-Van. "Understanding of Variables and Their Uses Acquired by Students in Traditional and Computer-Intensive Algebra." Doctoral dissertation, University of Maryland, 1990.

Palmiter, Jeanette R. "Effects of Computer Algebra Systems on Concept and Skill Acquisition in Calculus." *Journal for Research in Mathematics Education* 22 (March 1991): 151–56.

Pea, Roy D. "Beyond Amplification: Using the Computer to Reorganize Mental Functioning." *Educational Psychologist* 20 (4) (Fall 1985): 167–82.

———. "Cognitive Technologies for Mathematics Education." In *Cognitive Science and Mathematics Education*, edited by Alan Schoenfeld. Hillsdale, N.J.: Lawrence Erlbaum Associates, 1987.

Peschek, Werner, and Edith Schneider. "How to Identify Basic Knowledge and Basic Skills? Features of Modern General Education in Mathematics." *International Journal of Computer Algebra in Mathematics Education* 8 (1): 7–22.

Sfard, Anna. "On the Dual Nature of Mathematical Conceptions: Reflections on Processes and Objects as Different Sides of the Same Coin." *Educational Studies in Mathematics* 22 (February 1991): 1–36.

Sheets, Charlene. "Effects of Computer Learning and Problem-Solving Tools on the Development of Secondary School Students' Understanding of Mathematical Functions." Doctoral dissertation, University of Maryland, 1993.

CHAPTER 3

CAS as Pedagogical Tools for Teaching and Learning Mathematics

Bernhard Kutzler

COMPUTER algebra systems (CAS) encourage us to think about *what* mathematics we will teach and *how* we will teach it. In this chapter, I focus on the *how* of teaching mathematics and develop a two-level framework for understanding, categorizing, and planning the use of technology—in particular, CAS—in teaching and learning mathematics. I first consider *automation* and *compensation*, two types of support that a CAS tool can offer students and teachers. I next discuss four pedagogical approaches to teaching and learning mathematics: *trivialization, experimentation, visualization*, and *concentration*—all greatly facilitated by proper use of CAS. On the basis of the framework outlined above, I then introduce the *scaffolding method* as a pedagogically justified method of sequentially using and not using technology to achieve certain teaching goals.

Introduction

In mathematics education, computer algebra systems (CAS) have sparked a worldwide discussion regarding the value and "sensefulness" of what we teach in a traditional mathematics class. CAS help us recognize that we focus too much on teaching the craftsmanship of performing operations, skills easily performed by a computer. Hence, CAS encourage us to think about *what* we teach.

CAS can be used to teach mathematics in numerous ways—good ways and bad ways. Improper approaches to the use of CAS are often associated with "technology freaks," teachers who use CAS simply because the technology exists. CAS should never drive the mathematics we teach. Instead, our mathematics goals for teaching should drive the use of CAS. I agree with the statement of Helmut Heugl, the director of the Austrian Derive and TI-89/92 Projects, involving almost 6,000 students, "If it is not pedagogically justified to use CAS, it is pedagogically justified not to use CAS" (personal communication). Therefore, CAS also encourage us to think about *how* we teach.

Having previously written about the *what* of teaching mathematics (Kutzler 2002), I focus on the *how* of teaching mathematics by discussing the different roles that CAS tools can assume in teaching and learning mathematics. At the Austrian Center for Didactics of Computer Algebra, my colleagues and I advocate the importance of a pedagogically justified use of CAS—an approach that is driven solely by mathematics teaching goals. It is important to note that many of the ideas that follow are also applicable to tools other than CAS, such as graphing calculators.

Tools I: Automatization and Compensation

In a speech to educators in 1936, Albert Einstein made a remarkable analogy: "If a young man has trained his muscles and physical endurance by gymnastics and walking, he will later be fitted for every physical work. This is also analogous to the training of the mind and the exercising of the mental and manual skill."

On the basis of Einstein's analogy, I compare the discipline of *mathematics* as an intellectual achievement to the discipline of *moving-and-transportation* as a physical achievement—an idea that originated with Frank Demana (keynote lecture at ATCM [Asian Technology Conference in Mathematics], 1998, Tsukuba, Japan). The preceding analogy compares physical achievement to intellectual achievement, as both relate to the roles of technology.

The most elementary method of moving is *walking*. Walking is a physical achievement that requires mere muscle power. The corresponding activity in mathematics is *mental calculation*, which includes both mental arithmetic and mental algebra. Mental calculation requires mere brain power.

The process of *riding a bicycle* is a method of moving that uses a mechanical device—the bicycle—to make more effective use of the rider's muscle power. A person who is riding a bicycle can move greater distances or can move faster than a person who is walking. The corresponding activity in mathematics is *paper-and-pencil calculation*. Paper and pencil are used as an "external memory" allowing more efficient use of a person's brainpower.

Another method of moving is *driving a car*. The car is a device that produces movement. The driver needs very little muscle power to drive the car but must be able to start the engine, accelerate, steer, brake, and adhere to traffic regulations. The corresponding activity in mathematics is *calculator or computer calculation*. The calculator or computer produces the result, but its user needs to know how to operate it and how to push the correct buttons in individual situations.

How should we determine which method of moving is sensible for a specific situation? If I ask someone to obtain today's newspaper from a newsstand located 150 meters away, walking may be the best form of transportation. If the newsstand is 1000 meters away, riding a bicycle may be the most reasonable means of transportation. If the newsstand is 10 miles away, driving a car is probably advisable, particularly if time is a factor. In mathematics, the sensible use of technology is similar to the situation described above. The multiplication of 2 one-digit numbers is best performed mentally; the multiplication of 2 two-digit numbers is best accomplished by use of paper and pencil; the multiplication of 2 seven-digit numbers requires the use of a calculator.

Educators often assume that students who use a calculator to obtain the product of 7 and 9 lose the ability to perform mental arithmetic, an obvious example of improper use of technology that also occurs in subject areas other than mathematics education. Similarly, some people misuse their car by driving 150 meters to the next newsstand. Drivers who do so may harm themselves through a lack of physical exercise and may harm the environment through the unnecessary introduction of exhaust fumes. Despite the possible misuse of the car, we do not demand its abolition. Likewise, we should not banish calculators and computers simply because some students use them improperly. Just as we need to make people aware that physical exercise is essential for physical fitness and health, we need to make them aware that mental exercise—mental arithmetic and algebra—is essential for mental wellness and fitness. Trying to persuade students to do simple computations manually is usually unsuccessful, but by separating students' examinations into two parts—a compulsory part and a voluntary or "freestyle" part, we can encourage students to use manual computation when appropriate.

The preceding solution to students' improper use of technology is analogous to ice skating scores; two independent scores are added to determine the total score. In the compulsory part of a mathematics examination, we would primarily test mental skills and allow no technology—not even a four-function calculator. In the freestyle part of the same examination, we would test problem solving and allow students to use any kind of technology, CAS in particular, as described in publications by Kutzler (2000, 2002) and by Herget et al. (2000).

Continuing the analogy (see **table 3.1**), if the person who was asked to buy a newspaper from a newsstand 150 meter away is physically challenged or has a broken leg, walking 150 meters may be very difficult or impossible. A wheelchair is available to help people with physical limitations. *Using a wheelchair* is a method of technology-supported movement, with technology compensating for a person's physical weakness. But weaknesses also occur in intellectual activities. For example, a student with a weakness in solving systems of linear equations may find it very difficult, if not impossible, to solve analytic geometry problems. It is our pedagogical duty, an act of humanity, and a step toward "equity" to provide a mathematically challenged student with a tool that compensates for such a weakness. Only then does the student have a fair chance to succeed in analytic geometry despite his or her weakness.

Table 3.1
Analogy between transportation and mathematics

Moving and Transportation (Physical)	Mathematics (Intellectual)
Walking	Mental calculation
Riding a bicycle	Paper-and-pencil calculation
Driving a car	Calculator or computer calculation (automation)
Using a wheelchair	Calculator or computer calculation (compensation)

As I subsequently demonstrate, calculators and computers can be excellent mathematical compensation tools that allow mathematically challenged students to deal with advanced topics. The ultimate goal in teaching mathematics is to eliminate students' weaknesses in essential skill areas. According to Herget et al. (2000), the use of CAS forces us to reconsider what skills we consider essential. A physically challenged person may not be confined to a wheelchair for life if a physician can repair the patient's physical disability with individual therapy. Similarly, a teacher should attempt to repair a student's mathematical shortcomings with appropriate individual therapy. In both these situations, professionals are working to improve the "patient's" daily life—whether it involves shopping or doing analytic geometry—by providing an appropriate compensation tool, such as a wheelchair, a calculator, or a computer.

Automation and *compensation* are the two most elementary types of support that tools can provide. Although not specific to teaching and learning, *automation* and *compensation* are concepts that we can apply to CAS tools for teaching and learning mathematics. In general, CAS are more likely to serve as *automation* tools for "good" students and as *compensation* tools for "mathematically challenged" students. Because the terms "good" and "mathematically challenged" are relative to our teaching styles, their definitions will probably change when we begin to teach mathematics differently in terms of content and style.

Teachers have the pedagogical duty to use all available resources to facilitate the learning processes of their students. Students who are mathematically challenged—a group for which change is most urgent—seem to benefit from the use of compensation tools.

Tools II: Trivialization, Experimentation, Visualization, and Concentration

In this section, I discuss and demonstrate some specific roles of CAS as pedagogical tools by looking at four powerful approaches in mathematics teaching and learning: trivialization, experimentation, visualization, and concentration.

Trivialization

Referring again to my physical-intellectual analogy, the car broadened our moving-and-transportation horizons by trivializing movement for certain distances. Similarly, the calculator broadens our calculation horizons.

In the "old days" before scientific calculators, teachers carefully designed examination questions and homework problems so that all intermediate and final results were "nice." A "nice" result consisted of a solution that was an integer, a simple fraction, or a simple radical that might "disappear" later in the calculation process. Straightforward results were important so that students would not waste most of their time performing arithmetic operations. Because a scientific calculator can multiply 2 seven-digit numbers as quickly as 2 one-digit numbers, it has trivialized the process of performing arithmetic operations.

Drawing the graph of a linear function such as $y = 2x + 3$ is simple if one knows the geometric meaning of the two coefficients; only a "glimpse of talent," a pencil, and a straightedge are needed to produce a proper graph of the equation. Drawing the graph of a function such as $y = 2\sin(x/2) + \cos(x)$ is much more difficult, and producing a proper graph requires a reasonable degree of drawing talent. The graphing calculator can plot both functions (see **fig. 3.1**) in equal time, and it has no more difficulty plotting the second equation than plotting the first equation. Therefore, the graphing calculator trivializes the production of graphs.

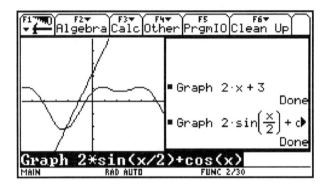

Fig. 3.1. Graph of function $y = 2x + 3$ and function $y = 2\sin(x/2) + \cos(x)$ using a TI-92 graphing calculator

Computing the first derivative of $y = x^n$ is simple if one knows the differentiation rule for powers. However, determining the first derivative of

$$y = \ln(\sin(\cos(\tan(\sqrt{x^2 - x + 1}))))$$

requires a lot of work, even by a good mathematician. The algebraic calculator can manage both equations above within seconds (see **fig. 3.2**). Therefore, the algebraic calculator trivializes algebraic symbolic computations. The trivialization of algebraic computation in mathematics teaching and the "white box/black box" principle were introduced in a landmark paper by Buchberger (1989).

$$\#2: \quad \frac{d}{dx} \text{LN}\left(\left|\text{SIN}(\text{COS}(\text{TAN}(\sqrt{x^2 - x + 1})))\right|\right)$$

$$\#3: \quad \frac{(1 - 2 \cdot x) \cdot \text{SIN}(\text{TAN}(\sqrt{x^2 - x + 1})) \cdot \text{COT}(\text{COS}(\text{TAN}(\sqrt{x^2 - x + 1})))}{2 \cdot \sqrt{x^2 - x + 1} \cdot \text{COS}(\sqrt{x^2 - x + 1})^2}$$

Fig. 3.2. Derivative of $y = \ln(\sin(\cos(\tan(\sqrt{x^2 - x + 1}))))$ using PC program Derive 5

Many tasks related to moving and transportation were once considered difficult but are routinely accomplished today by cars or other transportation devices. Similarly, CAS can be used to help students tackle problems that are more complex and more realistic, *one* aspect of using CAS that I consider the least important.

Experimentation

How did we discover all the mathematics we know today, and how do we discover more mathematics? According to an epistemologically oriented theory, we can visualize the main steps of mathematical discoveries as in **figure 3.3**. The process of applying known algorithms produces *examples*. From the examples we *observe* properties that are inductively expressed as a *conjecture*. Proof of the conjecture yields a *theorem*, that is, guaranteed knowledge. The theorem's algorithmically usable parts are *implemented* in a *new algorithm*. The algorithms are *applied* to new data, yielding *new examples*, which lead to new observations, new conjectures, and so on.

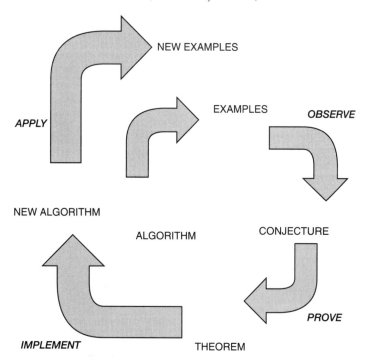

Fig. 3.3. *The main steps of mathematical discoveries*

The spiral proposed by Bruno Buchberger can be used to demonstrate the path of the discovery of mathematical knowledge (see **fig. 3.4**). A detailed description of Buchberger's creativity spiral and references to related models can be found in a publication by Heugl, Klinger, and Lechner (1996.) For my purposes in this chapter, I interpret Buchberger's creativity spiral *globally* as a part of the process of developing mathematical knowledge.

This creativity spiral can also be viewed as a visualization of the inductive concept-development process, in accordance with the constructivist approach of Piaget (1972), also reported by Marin and Benarroch (1994.) Buchberger's spiral contains three phases. During the *phase of experimentation*, known algorithms are applied to generate examples; a conjecture is then formed through observation. During the *phase of exactification*, the conjecture is turned into a theorem through proof; algorithmically useful knowledge is then implemented as a new algorithm. During the *phase of application*, algorithms are applied to real or fictitious data. In the application phase, solutions to *real problems* typically model real-world situations. Solutions to *fictitious problems* usually provide entertainment in the form of mind puzzles; they may also offer new knowledge that satisfies our "scientific curiosity."

The three phases of experimentation, exactification, and application—in both interpretations of Buchberger's creativity spiral—can be more elegantly denoted as *in*duction, *de*duction, and *pro*duction, respectively (see **fig. 3.4**). Although Piaget's concept-development process is inductive as a whole, it does contain inductive, deductive, and productive subprocesses.

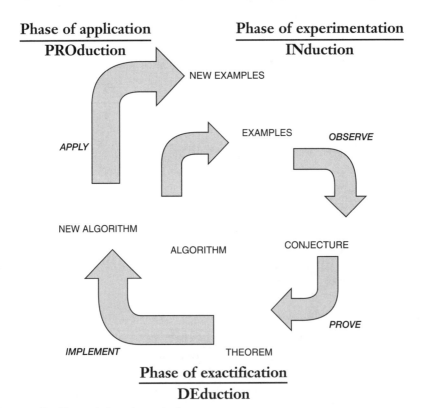

Fig. 3.4. Buchberger's learning spiral

Mathematics was originally an experimental science that consisted of the phases of *experimentation* and *application*. The Greeks later applied the deductive methods of their philosophy to mathematics and added the phase of *exactification*, thus establishing mathematics as the deductive science we know today. "Recently," in historical terms, a group of scholars—known by the name of Bourbaki—gathered around the French mathematician Dieudonne and restructured the mathematical knowledge using the system of definition-theorem-proof-corollary. The resulting *Bourbaki system*, developed for the purpose of innermathematical communication, includes only the phases of *exactification* and *application* and has become characteristic of modern mathematics. Bourbakism gradually lodged itself in teaching and learning. It has become customary to teach mathematics by deductively presenting mathematical knowledge and by asking students to learn and apply this knowledge to solving problems. But this approach is highly unnatural. Most of today's psychological theories of learning treat learning as an inductive process in which experimentation plays a pivotal role. In keeping with these theories, Freudenthal (1979) demanded that *"we should not teach students something that they could discover themselves."*

Few mathematicians could conduct mathematical research the way we demand that our students learn the subject of mathematics. A student must "locally" build an individual "house" of mathematics. A scientist must go through a similar process, one that is more global and operates on a much larger scale. For both the scientist and the student, a substantial part of knowledge acquisition occurs during the phase of *experimentation*. The reason that so many students are at odds with mathematics may be related to their lack of experimentation. Educators should demand that experimentation receives its due position in mathematics education. Phases of experimentation should complete or supplement the traditional teaching methods rather than act as substitutes for these methods! I am not suggesting that mathematics teachers return to Egyptian experimental mathematics but that they use all three phases of Buchberger's spiral in teaching their students.

However, within the framework of today's curricula, experimentation rarely occurs in our mathematics classrooms. Experimentation that is performed with paper and pencil is both time-consuming and prone to error. Using the time available in mathematics class, students can produce only a very few examples for the purpose of observing and discovering concepts. Furthermore, a "hefty" portion of student examples would probably be faulty due to calculation errors. Students cannot make valid observations and conclusions based on a few, partially correct examples! For example, in teaching geometry students that the three altitudes of a triangle—every triangle—intersect in one point, we might ask them to draw five triangles and construct the three altitudes in each triangle. Most students, being "lousy" draftspeople, find that the altitudes do *not* intersect in one point in *all* their five triangles. Therefore, they will probably not be convinced that the statement they were asked to illustrate is actually true.

CAS and interactive geometry software enable students to experiment with most topics taught in mathematics. Because these electronic assistants guarantee the "properness" of the results, students can complete several examples in a short time. Many great mathematicians have also used assistants. For example, Carl Friedrich Gauss employed a great number of "human calculators" who were instrumental in most of his famous discoveries. The great genius Johann Wolfgang Goethe called for "learning through doing and observing," an example of the phase of experimentation shown in the spiral (see **fig. 3.4**). Using CAS, we can now meet Goethe's demand.

Visualization

Visualization is defined as the illustration of an object, fact, or process with results that are graphic, numeric, or algebraic. Today, the term is primarily used for graphic illustrations of algebraic objects, numeric objects, or facts, describing either the process of illustration or the result of the illustration process. In countries that have widespread use of graphing calculators, visualization has become an important technique for teaching mathematics. Today visualization is primarily used to help students become competent at changing between representations, primarily for the purpose of studying the correspondence between algebraic and graphic representations. The example in **figure 3.5** shows how the parameter a affects the shape of the graph of the function $y = a \cdot x^2$.

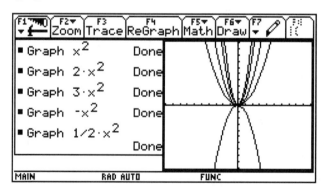

Fig. 3.5. *The effect of the parameter a on the graph of the function* $y = a \cdot x^2$

If we ask our students to manually draw the graphs of x^2, $2x^2$, $3x^2$, $\frac{1}{2}x^2$, and $-x^2$, the "good" students do fine, but for most of the "mathematically challenged" students, this assignment turns into a tedious graph-drawing exercise and they completely forget why they were asked to draw the graphs. Hence, the teaching goal is lost because of students' poor drawing skills. Using CAS or graphing calculators helps students with weak drawing skills stay connected with the goal of finding the relationship between the value of and the shape of the graph. Frank Demana and Bert Waits (1990, 1992, 1994) are the leading advocates of a teaching style that is called the *power of visualization* and is modeled by the preceding example.

Scientists who study the psychology of learning have discovered the concept of *reinforcement* and have shown that reinforcement works best if it immediately follows the action. A child who touches a hot stove can serve as an everyday, real-life example of reinforcement. The immediate pain that the child feels is the best prerequisite for the child's not touching the stove again. If the pain were felt several minutes later, the child probably would not connect the pain with the long-ago touching of the hot surface, nor would the child learn to avoid the stove.

When using a calculator as a visualization tool, immediate feedback is of central importance. If a student enters x^2 into the calculator and presses the proper key, the corresponding graph appears in a fraction of a second. The teacher can discuss the resulting picture immediately, and the student can associate the graph with the expression. Consider a mathematically challenged student, and compare the learning effect when the student must draw the graph manually with the effect when the student is allowed to use a calculator. A manually produced graph would be time-consuming and probably only vaguely resemble the true graph. Can a student make valid observations from an incorrectly drawn graph? Only with the help of the calculator does this student have a realistic chance of discovering and memorizing the correspondence between the expression and the correct graph. The process of producing graphs with paper and pencil certainly continues to be a worthwhile activity that is important for *understanding* the correspondence between algebraic and graphic representations. Yet, because immediacy and correctness are crucial psychological factors for mathematically challenged students, teachers should provide these students a computer algebra system or a graphing calculator.

The preceding situation very clearly demonstrates that the pedagogical goals connected with an activity determine *if* and *how* a teacher uses a tool to support the activity. As a result, the teacher has become more essential in a technology-supported mathematics environment than in a traditional mathematics-learning environment. Hence, the importance of preservice and in-service training for teachers increases. Although the power of visualization is commonly associated with graphing calculators, it is also an important approach for use with algebraic data—a concept that I demonstrate in the following section.

Concentration

We can compare teaching and learning mathematics to building a house, the "house of mathematics." Each mathematical topic that we teach forms a "story" or floor of the house. The first story of a house must be complete before the second story can be built. Similarly, the treatment of almost any mathematical topic requires the mastery of previously learned topics. I illustrate the preceding concept by considering the topic of solving a linear equation in one variable.

To solve for x in the equation $5x - 6 = 2x + 15$, the student must transform the equation into the form "$x =$" by applying an appropriate sequence of equivalence transformations. Typically, the student is advised to "bring terms containing x to one side of the equation, and bring all other terms to the other side." Therefore, the student may decide to first subtract $2x$:

$$(5x - 6 = 2x + 15) \qquad -2x$$

After choosing this equivalence transformation, the student must apply it to both sides of the equation and simplify:

$$3x - 6 = 15$$

Next the student must choose another equivalence transformation, for example $+6$:

$$(3x - 6 = 15) \qquad +6$$

The student must simplify again:

$$3x = 21$$

At this stage during the process of solving the equation above, the student might make the common mistake of choosing -3 as the equivalence transformation. Because we are interested in the practice of teaching mathematics, we want to know why the student made such an error. The student might offer the following argument: "There is a 3 in front of the variable x. To get rid of the 3, I need to subtract 3." Believing that the correct equivalence transformation was chosen, the student might have written this:

$$(3x = 21) \qquad -3$$

$$x = 18$$

What did the student do wrong, and how can technology help? An analysis of the equation above reveals two alternating tasks: (1) the choice of an equivalence transformation, and (2) the simplification of algebraic expressions. In the problem above, the choice of an equivalence transformation is a higher-level task because it is the essence of the strategy used to solve the equation; it is the *new* skill that the student must learn when solving equations. The simplification of algebraic expressions is a lower-level task that the teacher assumes the student has already mastered.

The diagram in **figure 3.6** demonstrates how a student, while trying to learn a new skill, must repeatedly interrupt the learning process to perform a calculation. If a chess player were repeatedly interrupted, like the student, during a difficult chess game, the interruption could influence the outcome of the game. Likewise, if the student makes a mistake while completing the lower-level task or calculation, the higher-level task might be severely affected, preventing the student from learning. The student incorrectly obtained $x = 18$ for the "solution" to the equation $5x - 6 = 2x + 15$. It appears that the student began the next line with "x = " *simply because the transformation −3 was chosen in order to generate "x =" on the left-hand side of the equation.* As a result, the student had the erroneous impression that −3 simplified the equation as desired, a misconception related to the higher-level skill. If the student had actually subtracted 3 from both sides of the equation, the resulting equation would not have been in the desired "$x =$" format.

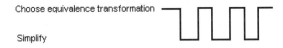

Fig. 3.6. The learning process when alternating a higher-level task with a lower-level task

A continuous "change of levels," as described above, inevitably occurs in almost all topics in school mathematics; it is one of the central problems in mathematics education because students must learn a *new* skill while practicing an *old* skill.

A CAS, such as Derive, could have been used as follows to facilitate the learning process of solving the equation $5x - 6 = 2x + 15$:

First the student enters the equation:

$$5x - 6 = 2x + 15 \text{ (ENTER)}$$

#1:   5·x − 6 = 2·x + 15

Next the student follows the input of the equivalence transformation:

$$\#1 - 2x \quad \text{(ENTER)},$$

then simplifies by clicking the button 🟰.

```
#2:     (5·x - 6 = 2·x + 15) - 2·x
#3:                                    3·x - 6 = 15
```

The simplification—application of the equivalence transformation to both sides of the equation—is performed by the CAS. Then the student chooses the next equivalence transformation:

$$\#3 + 6 \text{ (ENTER)},$$

then simplifies with $\boxed{=}$.

```
#4:     (3·x - 6 = 15) + 6
#5:                                    3·x = 21
```

If the student makes the mistake of choosing the equivalence transformation –3 to solve the equation, the screen below is displayed:

$$\#5 - 3 \text{ (ENTER)},$$

then the student simplifies with $\boxed{=}$.

```
#6:     (3·x = 21) - 3
#7:                                    3·x - 3 = 18
```

The CAS has obviously performed the simplification process properly. Because the CAS did not simplify the equation to resemble the expected "$x =$" form, the student receives immediate feedback that the transformation –3 was not successful.

By referring to the example above, we encounter two issues addressed previously in this chapter: (1) *experimental learning*, modeled by the student who experimented with possible equivalence transformations, and (2) *visualization*, illustrated by the immediacy of the results as the transformations were applied to the equation $5x - 6 = 2x + 15$.

A student who follows the CAS-supported exercise above can fully concentrate on the higher-level skill of choosing an equivalence transformation. The CAS performs the lower-level skill of simplification, at least for the moment (see **fig. 3.7**).

Fig. 3.7. The learning process when concentrating on the higher-level task

In previous publications (Kutzler 1998a,b), I explained in detail how to use Derive or the TI-92 and the TI-89 graphing calculators to teach the solution of linear equations. The algebraic approach illustrated above is discussed in further detail, as are numeric and graphic approaches.

The Scaffolding Method

In the preceding subsection of this chapter, I compared teaching mathematics to building a house. In the language of this house-building metaphor, as soon as teachers begin building the story of "choosing equivalence transformations," the story of "simplifying" is still incomplete for many students. Teachers simply do not have enough time to wait until each student has completed all previous stories of the house before adding the next story. The curriculum may force teachers to teach the next topic, independent of the progress of individual students. Therefore, we must ask ourselves "How can a student build a story of a mathematics house on top of an incomplete story?" (see **fig. 3.8**).

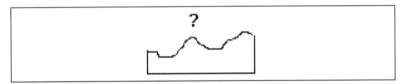

Fig. 3.8. For some students, the story of simplifying is still incomplete.

The diagram in **figure 3.9** illustrates my answer to the rhetorical question posed above. While the student learns the higher-level skill, the calculator solves all subproblems that require the lower-level skill. Using the language of the metaphor, the calculator is the scaffolding above the incomplete story of the house.

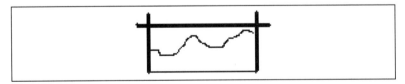

Fig. 3.9. Using scaffolding above the incomplete mathematical story

Having previously related the scaffolding method to solving a linear equation with Derive, I next apply the scaffolding method to solving a system of linear equations with Gaussian elimination. In attempting to solve for x and y in the system $2x + 3y = 4$, $3x - 4y = 5$, the student enters the equations:

$$2x + 3y = 4 \quad \text{(ENTER)}$$

$$3x - 4y = 5 \quad \text{(ENTER)}$$

```
#1:     2·x + 3·y = 4
#2:     3·x - 4·y = 5
```

Gaussian elimination requires the student to choose a linear combination of the two equations, so one variable is eliminated—the new skill that the student must learn. The teacher assumes that the student has mastered the prerequisites of simplification, substitution, and solving an equation in one variable. First the student tries to eliminate y by finding the sum of four times the first equation and three times the second equation:

$$4*\#1+3*\#2 \text{ (ENTER)}$$

then the student simplifies with [=].

```
#3:     4·(2·x + 3·y = 4) + 3·(3·x - 4·y = 5)
#4:                                              17·x = 31
```

Voila! Variable y disappeared, as requested. Typically the topic of solving of a system of linear equations with Gaussian elimination is taught in a paper-and-pencil environment in school. Some students may choose the correct linear combination, but because of a calculation error, the variable does not disappear. Other students may choose an incorrect linear combination, but the variable disappears because it "must" disappear. For both groups of students, their weakness in algebraic simplification are stumbling blocks (see **fig. 3.10**) to successfully learning the basic technique of Gaussian elimination; students in both groups continue to lag behind as they move "higher" in the house of mathematics. In previous publications (Kutzler 1998c,d), I have described ways to use Derive or the TI-92 or TI-89 graphing calculator to teach systems of linear equations and have discussed numeric methods, graphic methods, and the substitution method.

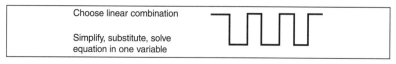

Fig. 3.10. *The learning path for Gaussian elimination*

In the exercise above, the algebraic calculator acts as a scaffolding that compensates for weaknesses associated with the lower-level skills; hence it helps students avoid mistakes. CAS can be used as pedagogical tools in a variety of ways. If the final teaching goal is for students to solve systems of equations manually (or perform any other manual skill labeled "B" on the diagram in **figure 3.11**) for an end-of-year assessment test), then I recommend the following three steps: (1) Teach skill A, and require students to practice it manually. (2) Teach skill B, and require students to practice it manually while allowing them to use the algebraic calculator to solve subproblems that require skill A, thus allowing students to focus on skill B. (3) Combine skills A and B, but do not allow students to use technology for support.

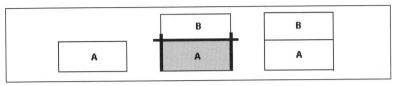

Fig. 3.11. Three-step technologically aided process for teaching students to a manual skill

The *scaffolding method* is any pedagogically justified sequence of using and not using technology for trivialization, experimentation, visualization, or concentration in the sense of either automation or compensation. The temporary use of technology can help students separate the learning process into smaller, more "digestible" pieces. For students who could not swallow the "big" pieces that we offered them in traditional mathematics teaching, the scaffolding method may help them master the necessary learning steps; it helps students keep track of the steps without getting lost in details such as simplification. When comparing the teaching of mathematics to building a house, the use of technology is the scaffolding.

Technology can and should be introduced as a pedagogical tool independent of changes to the curriculum or the assessment scheme. Technology can be used successfully by teachers who allow technology to be used during examinations and by those who do not. The scaffolding method is designed to support the learning process and traditional teaching goals while helping students acquire intellectual and mathematical skills. Consequently, CAS can help learners at all levels of mathematics in secondary education.

Conclusion

In 1991, all general and technical high schools in Austriawere equipped with the Derive computer algebra system. In response, a research project was conducted that became known as the Austrian Derive Project, a program involving 800 students who were taught regular mathematics with Derive. The results of the investigations described above were published by Heugl, Klinger, and Lechner (1996) in a German publication and also published in English by Aspetsberger and Fuchs (1996.) In the academic year 1997–1998, the "Austrian TI-92 Project I" was conducted, a program involving 2,000 students who used TI-92 calculators in regular mathematics classes; the "Austrian TI-92 Project II" followed with 3,000 students. The results of the Austrian investigations and other investigations, available on the Internet at the address http://www.acdca.ac.at, showed that if technology is used properly, it leads to these benefits:

- More efficient teaching and learning
- More independent productive student activity
- More student creativity
- Increased importance of the role of the teacher

The teacher has the duty to accompany and direct individual students on the voyage of discovery through the world of mathematics. Consequently, success in teaching mathematics is a result of good teachers who have received high quality teacher training. No technology exists that will change teaching, but technology can be a catalyst for teachers to change their teaching methods and focus as they aim toward teaching mathematics better.

REFERENCES

Aspetsberger, Klaus, and Karl Fuchs, guest eds. *The International DERIVE Journal* 3(1) (1996): 1–106.

Buchberger, Bruno. *Why Should Students Learn Integration Rules?* RISC-Linz Technical Report no. 89-7.0. Austria: University of Linz, 1989.

Demana, Franklin, and Bert K. Waits. "Implementing the Standards: The Role of Technology in Teaching Mathematics." *Mathematics Teacher* 82(1) (January 1990):27–31.

———. "Soundoff: A Computer for *All* Students." *Mathematics Teacher* 84(2) (February 1992): 94–95.

———. "Graphing Calculator Intensive Calculus: A First Step in Calculus Reform for All Students." In *Proceedings of the Preparing for New Calculus Conference*, edited by A. Solow. Washington, D.C.: Mathematical Association of America, 1994.

Einstein, Albert. *Out of My Later Years.* n.p.: Carol Publishing Co., 1956.

Freudenthal, Hans. *Mathematik als pädagogische Aufgabe.* Stuttgart: Klett Studienbücher, 1979.

Herget, Wilfried, Helmut Heugl, Bernhard Kutzler, and Eberhard Lehmann. "Indispensable Manual Calculation Skills in a CAS Environment." *Ohio Journal of School Mathematics* 42 (Autumn 2000): 13–20. Also published in *Exam Questions and Basic Skills in Technology-Supported Mathematics Teaching (Proceedings Portoroz 2000)*, edited by V. Kokol-Voljc et al., pp. 13–26

(Portoroz, Slovenia: 6th ACDCA Summer Academy, 2–5 July 2000). Also published in Micromath 16(3) (2000): 8–12.

Heugl, Helmut, Walter Klinger, and Joseph Lechner. *Mathematikunterricht mit Computeralgebra-Systemen (Ein didaktisches Lehrerbuch mit Erfahrungen aus dem österreichischen DERIVE-Projekt)*. Bonn: Addison-Wesley Publishing Co., 1996.

Kutzler, Bernhard. *Improving Mathematics Teaching with DERIVE*. Bromley: Chartwell-Bratt, 1995.

———. *Introduction to the TI-92 (Handheld Computer Algebra)*. Hagenberg: BK Teachware, 1996.

———. *Solving Linear Equations with Derive (Experimental Learning/Visualization/Scaffolding Method)*. Hagenberg: BK Teachware, 1998a.

———. *Solving Linear Equations with the TI-92 (Experimental Learning/Visualization/Scaffolding Method)*. Hagenberg: BK Teachware, 1998b.

———. *Solving Systems of Equations with Derive (Experimental Learning/Visualization/Scaffolding Method)*. Hagenberg: BK Teachware, 1998c.

———. *Solving Systems of Equations with the TI-92 (Experimental Learning/Visualization/Scaffolding Method)*. Hagenberg: BK Teachware, 1998d.

———. "Two-Tier Exams as a Way to Let Technology In." In *Exam Questions and Basic Skills in Technology-Supported Mathematics Teaching (Proceedings Portoroz 2000)*, edited by V. Kokol-Voljc et al., pp. 121–24. Portoroz, Slovenia: 6th ACDCA Summer Academy, 2–5 July 2000.

———. "What Math Should We Teach When We Teach Math with CAS?" (in Italian). In *Proceedings of the 4th Covegno Nazionale de ADT (Associazione per la Didattic con le Technologie)*, pp. 134–40. Monopoli, 11–13 October 2002.

Marin, N., and A. Benarroch. "A Comparative Study of Pagetian and Constructivist Work on Conceptions Science." *International Journal of Science Education* 16(1) (1994): 1–15.

Piaget, J. *Die Entwicklung des Erkennens I – Das mathematische Denken*. Stuttgart: Klett, 1972.

Activity 2

Let $f(x)$ be a cubic equation with three real roots, a, b and c, and let $m = \frac{1}{2}(a + b)$ the average of any two roots. Prove that the tangent line to $f(x)$ at $x = m$ intersects the cubic at $(c, 0)$, the third root of the cubic. (Example suggested by Jim Schultz)

Solution:

- $(x-a)(x-b)(x-c) \to f(x)$ Done

- $\dfrac{a+b}{2} \to m$ $\dfrac{a+b}{2}$

- solve $\left(\left(\dfrac{d}{dx}(f(x))\Big| x = m\right)\cdot(x-m)+f(x)=0, x\right)$

$$x = c \text{ or } a^2 - 2ab + b^2 = 0$$

The last line finds the x-intercept of the tangent line at $x = m$. The x-intercept is $x = c$ or $(c, 0)$. QED.

Note: The solution by CAS draws our attention to the case where exactly two roots are equal $a = b$, that is, where $a^2 - 2ab + b^2 = (a - b)^2 = 0$. In this case, the tangent line must be horizontal; therefore, the tangent line (the x-axis) contains the other root. In this instance, the CAS actually helps us do more in-depth mathematics.

Alternative solution:

- $\left(\dfrac{d}{dx}(f(x))\Big| x = m\right)\cdot(x-m)+f(m) \to t, x$ Done

- $y = \text{factor}(t(x))$ $y = \dfrac{^-(a-b)^2\cdot(x-c)}{4}$

The factored form of the tangent line,

$$t(x) = -\dfrac{(a-b)^2}{4}(x-c),$$

also shows that the x-intercept is $(c, 0)$.

Chapter 4

Thinking out of the Box

William G. McCallum

STUDENTS, parents, and teachers often think of technology as a black box that performs calculations in an unknown way, the results of which are available for more studied use. Some people value the black box for its ability to skip mindless drudgery and get to the concepts; others fear that it replaces the mental exercise that is necessary preparation for deeper understanding. Some educators seek to avoid this problem by making the black box a white box, still taking advantage of technology instead of banning it. A white box is defined as a transparent device that shows each step in the calculation, although the steps themselves are carried out by machine. Whatever the merits of technology, the box metaphor has the following bad effects:

- Students treat the boxes, black or white, as oracles: "The answer must be right because my calculator said so." This assumption permits students to avoid responsibility for their answers.

- Teachers treat the boxes as miraculous timesaving devices: "I won't have to teach that because they'll do it with the calculator." This assumption is dubious. Although a chain saw is more powerful than an axe, it takes just as long to learn how to use a chain saw.

- Critics of technology seem to think that the boxes have strange, mind-sapping powers: "When the calculator turns on, the brain turns off." This attitude deflects attention from other, less magical causes for mindless behavior, such as popular attitudes toward the nature of mathematics and toward intellectual endeavor in general.

Although a grain of truth is apparent in each of these beliefs, the metaphor of the black box amplifies them to an irrational degree, perhaps because of its resonance with the myth of Pandora's box or with that other real but problematic box in our lives, the television, to which is attributed the same array of oracular, miraculous, and demonic abilities.

Beyond the Box Metaphor

The undesirable consequence of accepting the black-box metaphor is that the black box, not the mathematics, becomes the center of attention. Thinking of the technology as a white box whose internal operations are visible is a helpful first step toward refocusing on the mathematics rather than the technology. However, I would like to suggest that, for the purposes of considering its uses in teaching, we should let technology out of its metaphorical box entirely so it can take its place in the general landscape beside the older miracles of paper and pencil. Indeed, the power of computer algebra systems (CAS) is no longer confined to black boxes but is freely available on the World Wide Web in such sites as http://www.integrals.com. Immense computational power is in the air that we and our students breathe. In our daily lives, we accept and adjust to new technological miracles; we welcome their advantages, recognize their dangers, and ultimately adjust to them, forget about them, and continue with the important business of living. It is a good idea to do the same in our teaching lives.

Putting the Mathematics First

To consider sensibly the uses of technology, we must, paradoxically, start out by forgetting about it. In the same way that we might plan a trip by first choosing a destination on a map and then considering the details of getting there, we should start out by first considering our mathematical goals and then considering the ways we and our students can get there. Furthermore, in planning a trip for our children, we consider not only the means of travel we prefer but also the ones they prefer. Similarly, in planning how to teach our students, we should consider the new technologies they use and the old technologies of paper and pencil that we used growing up.

Thus, I propose the following procedure for thinking about using CAS in teaching:

Step 1. Think of a piece of mathematics you want your students to learn: a calculation, a concept, or a connection.

Step 2. Formulate a question to test this knowledge. Put yourself in your students' shoes, and solve using the various technologies available—including paper and pencil and computer algebra systems. Next ask a student to work the problem.

Step 3. Evaluate the mathematical content of the solution, revise step 1, and repeat.

No doubt, this advice is unremarkable. However, it is worth making explicit because CAS are treated like the new kitchen gadget we just bought and do not quite know what to do with. We are still in those first few days when we keep trying to think of ways to use or avoid using the gadget. The more sensible approach is to decide what we want to eat first, prepare it using what comes to hand, and see if we like the result. A willingness to try new technologies as well as old, coupled with good taste and common sense in evaluating the results, ensures that the new gadget finds its niche.

I illustrate the process with a very simple dish. Many examples of interesting, creative problems that use the full power of a CAS have been created. However, if the CAS is used only in specific preparations—the pasta machine of mathematics education rather than the food processor—then perhaps we do not need a whole book devoted to its use. I have chosen, more or less at random, a fairly simple example to work on and will perform four "rounds" of the process. Perhaps the most surprising result is the extent to which step 2 above forces us to focus on underlying *mathematical* issues as we consider all the *technologies* available to our students

Round 1: The Distinction between Technique and Insight

Topic for the day: solving equations

To determine whether our students can take an equation and put it in a form amenable to solution using the quadratic formula, we formulate the following problem.

PROBLEM 1. Solve the equations

$$x + \frac{1}{x} = 0, \quad x + \frac{1}{x} = 2, \quad x + \frac{1}{x} = 4.$$

Both paper-and-pencil solutions and CAS solutions are shown subsequently, along with our comments.

PAPER-AND-PENCIL SOLUTION

$x + 1/x = 0$ $\qquad$ $x + 1/x = 2$ $\qquad$ $x + 1/x = 4$

$x^2 + 1 = 0$ $\qquad$ $x^2 + 1 = 2x$ $\qquad$ $x^2 + 1 = 4x$

$x = \pm\sqrt{-1}$ $\qquad$ $x^2 - 2x + 1 = 0$ $\qquad$ $x^2 - 4x + 1 = 0$

$\boxed{\text{No Solution}}$ $\qquad$ $(x-1)^2 = 0$ $\qquad$ $x = \dfrac{^-(-4) \pm \sqrt{(^-4)^2 - 4 \cdot 1 \cdot 1}}{2}$

$\qquad\qquad\qquad\qquad\quad$ $\boxed{x = 1}$ $\qquad$ $\boxed{x = 2 + \sqrt{3} \text{ or } x = 2 - \sqrt{3}}$

Good problem!

(a) Tests ability to put equation into standard form
(b) Tests ability to choose the appropriate method: direct observation, factoring, or the quadratic formula

■

CAS SOLUTION

solve(x+1/x=0, x)

$\qquad\qquad\qquad\qquad$ False

solve(x+1/x=2, x)

$\qquad\qquad\qquad\qquad$ x = 1

solve(x+1/x=4, x)

$\qquad\qquad\qquad\qquad$ x = −(√3 − 2) or x = √3 + 2

What?

Requires barely any effort beyond pushing a few buttons; no evidence that the student understands anything

Ban the black box!

■

Thinking out of the Box

The students' marginal comments might be quite different from ours. A natural reaction of students would be to question why they are being required to learn paper-and-pencil solutions when the calculator does such a good job. We must think of some good responses to this question.

After seeing the CAS answers to the foregoing problems, we worry that students who use CAS are learning nothing more than how to operate a machine. The machine does replace much of what we used to teach, but will students learn to exercise any judgment in using it? Will they know how to evaluate their solutions and recognize when their answers do not make sense? Of utmost importance, will they learn to recognize situations where they do not need to use CAS? For example, we immediately recognize that $x + 1/x = 0$ cannot have a solution; if x and $1/x$ are both defined, then they are both nonzero and of the same sign.

Then the thought occurs to us, Should the same questions that we asked previously regarding students using CAS be asked about the students who gave the paper-and-pencil answer? With newfound concerns about what is going on inside the students' heads, we realize that although the students' solutions demonstrate the abilities (a) and (b) that we noted in our comments, they possibly represent reflexive behaviors rather than instances of conscious judgment. Is it possible that the student giving the first answer is functioning somewhat like the CAS in the second answer?

Round 1: Back to the Drawing Board

Returning to step 1 of the procedures for using CAS in teaching, we need to think again about what we really want students to know. We remember that we chose the equations because we expected students to recognize that they were quadratic equations in disguise. We want students to look at an equation and know what moves can be made with it, just as a good chess player can look at a board and see potential traps and opportunities. For example, we want students to anticipate the degree of the transformed equation and then predict the number of solutions. This goal encourages us to formulate a different problem using the same equations:

PROBLEM 2. For what values of a does

$$x + \frac{1}{x} = a$$

have no solutions? Exactly one solution? More than one solution?

We subsequently return to the solution to this problem, but we first pause to reflect.

What have we learned from round 1?

Our consideration of the CAS answer to our problem has forced us to acknowledge an explicit teaching goal: *algebraic insight*, the ability to discern the structure of an algebraic expression and the potential forms inherent in it. This goal is purely mathematical, not dependent on technology. However, before the CAS appeared on the scene, it was easy to assume—a dubious assumption—that it was a concomitant of technical skill and, therefore, to ignore it.

The appearance of CAS not only forces us to pay attention to algebraic insight, it also forces us to reconsider the role of paper-and-pencil algebraic technique. Training in algebraic technique may remain good preparation for the development of algebraic insight, but we can no longer offer our students the thought that motivated me as a child. I was sure that by learning to solve equations, I was learning the way into a universe that other people could not enter if they could not follow me. The mere ability to carry out the steps in solving problem 1 above is no longer the only key to the magic kingdom. Apparently, another entrance to this magic world exists. Even those educators who think that entering the kingdom using technology is a false and easy entrance have a responsibility to ensure that students who use it arrive in good shape. We cannot stop students from using CAS any more than we can stop them from breathing.

Round 2: Brave New World

Returning to the solutions to problem 2, we consider the implications.

PAPER-AND-PENCIL SOLUTION

$$x + 1/x = a$$
$$x^2 + 1 = ax$$
$$x^2 - ax + 1 = 0$$
$$-ax = -x^2 - 1$$
$$a = \frac{-x^2 - 1}{-x} = \frac{x^2 + 1}{x}$$

??

∎

FIRST CAS SOLUTION

```
solve(x+1/x=a, x)
```

$$x = \frac{-(\sqrt{a^2 - 4} - a)}{2} \text{ or } x = \frac{\sqrt{a^2 - 4} + a}{2}$$

If $a^2 > 4$, the expression under the square root sign is positive; therefore the equation has two roots. If $a^2 < 4$, the expression is negative; therefore, the equation has no roots.

SECOND CAS SOLUTION

solve(x+1/x=0,x)

$$\text{false}$$

solve(x+1/x=2,x)

$$x = 1$$

solve(x+1/x=4,x)

$$x = -(\sqrt{3} - 2) \text{ or } x = \sqrt{3} + 2$$

If $a < 2$, the equation has no roots; if $a = 2$, it has one root; if $a > 2$, it has two roots. ■

THIRD CAS SOLUTION

factor(x+1/x−a)

$$\frac{x^2 - a \cdot x + 1}{x}$$

Multiplying by x, we see that we must solve a quadratic equation. The quadratic formula gives

$$x = \frac{a \pm \sqrt{a^2 - 4}}{2}.$$

If $a > 2$ or $a < -2$, then the expression under the square root sign is positive, so the equation has two distinct roots. If $a = \pm 2$, it is zero, so the equation has only one root. If $-2 < a < 2$, the expression is negative, so the equation has no real roots. Also, since $a^2 - 4 \uparrow a^2$, none of the roots is ever zero, so no extraneous roots result from multiplication by x. ■

At this stage we begin to regret our pious words about letting technology out of the box. What were we thinking? The student using paper and pencil has been completely thrown off guard by the introduction of the parameter a. Although this student can transform algebraic equations accurately, *knowing which variable to solve for* seems to be difficult.

The first CAS student is in better shape—probably a result of good luck. Entering the same command as before, but with an *a* in place of the 0, 2, or 4, the student has been presented with the solution by the CAS. Yet this student manages to do something with the output. A significant new step has appeared in this problem: recognizing for which values of *a* the square root $\sqrt{a^2-4}$ exists and for which values it is zero. However, the solution illustrates a problem with the CAS. The CAS has presented the two roots in a slightly nonstandard form, causing the student difficulty in detecting the case when they are equal, and the student has obviously missed that case. Also, the student's division into cases $a^2 > 4$ and $a^2 < 4$ is not entirely satisfactory—possibly due to a lack of understanding exactly what ranges of *a* these inequalities define.

The second CAS student, not realizing that the equation with the parameter can be put into the CAS, has just reentered the three equations from problem 1 and made a guess based on them. This student has forgotten that the value of *a* may be negative and has given no real mathematical justification for the answer. However, this student has shown something that neither the student using paper and pencil nor the first student using CAS demonstrated. The second student using CAS has demonstrated an understanding that the problem represents a family of equations parameterized by *a* and considers their solutions in *x* for different values of *a*.

Obviously one of the parents of the third CAS student must be a mathematics teacher. This kid is brilliant! This student not only knows the quadratic formula but also realizes the need to eliminate the possibility of a zero root and additionally has discovered that the CAS does the algebra if one puts an equation in the form *something* = 0 and then factors *something*. This third student using CAS demonstrates the type of judgment we want our students to possess—clever use of technology coupled with masterful insight.

We return once again to step 1 of the procedures for using CAS in teaching, this time wondering whether any of the students other than the third student using CAS have a solid grasp of what an equation is. To test students' conceptual understanding of an equation, we formulate the next question.

PROBLEM 3. Do three distinct numbers *a, b, c* exist such that

$$a+\frac{1}{a}=b+\frac{1}{b}=c+\frac{1}{c}?$$

To answer this question, the student must recognize that it can be reformulated as follows: does a number *d* exist such that the equation $x+1/x = d$

has three distinct solutions? We *are* requiring students to supply the variable and to introduce a new parameter. How do they handle this question?

What have we learned from round 2?

First, we have learned that creating good questions is a difficult task. We no longer know what types of answers, right or wrong, to expect. Students will use the CAS in ways we had not anticipated; they will miss the point we had intended in asking the question but perhaps will stumble onto another point we had not thought of making. We must learn to expect a wider variety of answers and be more careful in interpreting them.

Second, CAS syntax both helps and hinders students. On the one hand, the syntax required by the **solve** command forces students to state explicitly, at the outset, which variable is being solved for, thus preventing the CAS student from getting "lost" in the same way the student using paper and pencil became confused. On the other hand, the output of the CAS may appear in an unanticipated form, making the task more difficult for students.

Third, we have learned that we cannot go back. We have just discovered something about the fragility of the mathematical understanding of the student using pencil and paper. We cannot ignore that revelation—even if the CAS disappears tomorrow. The paper-and-pencil exercises shown in this chapter were designed to form a complete picture so that students will put it all together and develop some independent thinking. Unfortunately, whenever we vary the wording of our question slightly or introduce a seemingly harmless parameter, we derail students completely. Although the CAS has arrived in mathematics education, plenty of work still remains for teachers—that is good news. Contrary to our fears, the CAS does not provide students a magic way of avoiding thought. Indeed, the only students who truly master the CAS seem to be those who would have thrived without it. For the other students, perhaps they will experience as many learning opportunities with the CAS as without it. With or without CAS, we must work harder at affording our students those opportunities.

Round 3: Back to Basics

We have been waiting eagerly for the answers to problem 3, but all the pages are blank. Students do not know what to make of this question—a pity, it was such a good question. In the process of reformulating the original problem to take account of the original CAS solution, we have also made it much more difficult. What started out as a simple exercise has become a test of sophisticated and flexible algebraic knowledge.

Do we still need problems like problem 1 to build the intuition needed for problems 2 and 3? Must we insist that they be done by hand? We return to step 1 of the procedures for using CAS in teaching with a renewed interest in the connection between mechanical algebraic skills and algebraic insight. We find ourselves wandering into unknown territory. Those of us brought up in the pre-CAS era have difficulty imagining how students can develop the ability to recognize that $x + 1/x = a$ can be transformed into a quadratic equation if they have not spent many hours working paper-and-pencil exercises. We possess this ability and insight because we can move the symbols as rapidly in our heads as we have so often moved them on paper, performing a sort of fast-forward computation and keeping track of certain aspects of the equation, such as its degree. I do not know if it is possible to develop these abilities without paper-and-pencil skills, but it seems safe to suppose that at least some paper-and-pencil drill will still be necessary.

However, elementary exercises with the CAS are probably necessary if we want students to use it with flexibility and thoughtfulness later on. The answers we have seen so far suggest a number of possibilities. First, the clever use of the **factor** command, which our bright student discovered early on, should perhaps be added to the library of standard techniques that we teach our students. Second, the CAS students were aided by the fact that the **solve** command takes as an argument the variable to be solved for. Perhaps we can use this feature of the CAS to develop exercises to test students' ability to decide what to do when given a choice of variables.

Here is a problem that aims to exercise both paper and pencil and CAS skills.

PROBLEM 4. Given a, for what values of x does the following equation hold?

$$x^2 - ax + 1 = 0$$

Given x, for what values of a does it hold?

In problem 4, I avoided the instructions "Solve for x" and "Solve for a." It is quite possible for a student to be adept at performing the paper-and-pencil activity triggered by these two instructions but be unable to articulate which variable is being asked for when. Similarly, the student with the CAS is challenged to enter the correct command:

$$\text{solve}(x^2 - a*x + 1 = 0, x) \text{ or } \text{solve}(x^2 - a*x + 1 = 0, a)$$

What have we learned from round 3?

We have learned that there are many questions we cannot answer. CAS have made us think more about algebraic insight: the ability to recognize algebraic structure, an understanding of the meaning of equations, an appreciation of algebraic manipulations as steps in a process of reasoning. However, students must have practice with basic procedures to develop these higher-level skills. I have not provided a solution to the problem in this round because I think that at this stage of our inquiry, we have passed the limits of armchair speculation. We need to decide which of the traditional drill questions still serve a purpose in developing algebraic insight and which were merely training our students to be good human algebra systems. The same decision goes for the CAS. We must develop drill exercises for the CAS that support the development of algebraic insight. Mathematics education research has something to say about attaining this goal, as do teachers experimenting with CAS in their classrooms every day. We recognize in the course of this experimentation, however, that our goals have fundamentally changed. Efficient and rapid computational ability is no longer an end in itself; if rapid calculation is all we want, we can get it from the technology. The technology is a means to a higher end, possibly many higher ends. We need to put more thought into defining those higher ends and develop more practical experience regarding the drill exercises that support these goals.

Round 4: Getting Thinking Out of the Black Box

The ease with which students can solve equations on the CAS opens up another category of problems that was not feasible before. If we play with the CAS and find solutions to $x + 1/x = a$ for various values of a, we notice patterns:

solve(x+1/x=4,x)

$$x = -(\sqrt{3} - 2) \text{ or } x = (\sqrt{3} + 2)$$

solve(x+1/x=5,x)

$$x = \frac{\sqrt{21} + 5}{2} \text{ or } x = \frac{-(\sqrt{21} - 5)}{2}$$

solve(x+1/x=6,x)

$$x = -(2 \cdot \sqrt{2} - 3) \text{ or } x = 2 \cdot \sqrt{2} + 3$$

The rational part of each root, the summand not involving a square root, seems always to be $a/2$, a simple consequence of the quadratic formula. Can we turn the problem around: encourage the students to see the pattern, try to explain it, and ultimately derive the quadratic formula in this case? The following problem is a move in this direction. Problem 5 can be assigned as an extended project, requiring repeated attempts and guidance from the teacher.

PROBLEM 5. Consider the equation

$$x + \frac{1}{x} = a.$$

1. Solve the equation for various values of the constant a. Find the sum and the product of the roots, and describe any patterns you notice.

2. Suppose that r is a root of the equation. Explain why $1/r$ must also be a root. Use this fact to explain the patterns you noticed in part 1.

Call the two roots of the equation r and s. By the end of the solution to the problem above, the students have discovered a proof that $r + s = a$ and that $rs = 1$. They can follow several different directions. One possibility is to make the connection with the related quadratic equation and its factored form:

$$x^2 - ax + 1 = (x-r)(x-s) = x^2 - (r+s)x + rs$$

Alternatively, students first generalize the project to the equation

$$x + \frac{b}{x} = a.$$

Playing around with the solutions to this equation using a CAS, then repeating the steps of the problem, again leads to a natural conjecture and proof. In this case the related quadratic equation

$$x^2 - ax + b = 0$$

has a quite general, although nonstandard, form.

Finally, students may extend this project into a derivation of the quadratic formula, either for the special equation $x^2 - ax + 1 = 0$ or for the more general form $x^2 - ax + b = 0$. Having found a formula for the sum and product of the roots, students are led to the following question: if we know the sum and the product of two numbers, do we know the numbers themselves? An opportunity for exploration with a CAS presents itself. If we can use CAS to solve simultaneous nonlinear equations, then we can use it to solve $x + y = a$, $xy = b$ for various values of a and b and try to see patterns. This investigation, combined with the previous extended project, would provide a proof of the quadratic formula.

What have we learned from round 4?

Many educators believe that the quadratic formula is no longer necessary to teach because students can use calculators to find roots, either numerically or exactly. On the one hand, if students are interested only in finding roots, then they can use the CAS. On the other hand, the derivation of the quadratic formula is a beautiful piece of mathematics, one that is often regarded as beyond the reach of the typical high school algebra class. Paradoxically, although the CAS buries the quadratic formula as algorithm, it resurrects the possibility of the quadratic formula as mathematical reasoning because it replaces the procedure of assertion-followed-by-derivation with observation-followed-by-explanation. This strategy makes an enormous difference in students' learning. For students who have not acquired the algebraic insight to follow one, a derivation is a leap in the dark that may end up being a leap of faith rather than a leap of understanding. Lost in the middle of a derivation and heading toward a conclusion, students may well conclude that mathematics is a meaningless game and that they are not good players.

However, if students start with the conclusion—obtained through direct observation—then the situation is quite different. Students cannot proceed to the step of explaining the observation until it has been clearly formulated, because students themselves have the responsibility for making the formulation. Therefore, students have a better chance of understanding the statement they are trying to derive. The process of explaining it is no longer a game but an act of scientific inquiry. The difference is the same as that between studying the life of Sherlock Holmes and the life of Shakespeare or between studying the anatomy of a dragon and the anatomy of a horse. Students use technology to remain engaged with mathematical reality instead of making things up as they go along.

Conclusion

The existence of symbolic computational capacity outside the human brain is something fundamentally new in our world; it is likely to change the mathematical life of students in our classrooms in the same way that the automobile changed the physical life of an earlier generation. To ensure that change for the better outweighs change for the worse, we need to first accept this fundamentally new thing and become knowledgeable about its capabilities but then relegate it to the background while we focus on our mathematical goals. One way of doing so, which I have sketched in some detail in this chapter, is to perform thought experiments with specific mathematical problems and see how they work it in this new technology-saturated environment. I conclude with three questions that have emerged from this thought experiment.

What is the role of mental exercise in developing mathematical ability?

It is commonly recognized that physical exercise is necessary for health; people with cars bicycle to work, and people who drive their cars to work make time in the day to engage in physical activity with no useful purpose other than the exercise it gives the body. By the same token, although many basic mathematical operations can now be performed faster and more accurately outside the human brain than inside, it seems likely that for the sake if its own mental health, the human brain will continue to need to perform some of those operations as exercises. The question is, What type of exercises? More important, how do we present these exercises to our students and convince them of their benefits? The old standby, "You have to know how to do this or you won't be able to get by in life," is no longer convincing, because students see technology as a way of getting by in life without the mental operations it replaces.

What technological skills are necessary for developing mathematical ability?

Just as we want our students to be safe and attentive drivers, we want them to use technology thoughtfully. To find out what constitutes a thoughtful use of technology, we need to use it thoughtfully ourselves. By picking up a CAS and working through problems with it, we can begin to lay down "rules of the road" for using it. It is important to temporarily shut out of our minds the back-seat driver who keeps reminding us of the way all mathematics was done in the past or could be done without the technology today. The creative use of the factor command in the third CAS solution to problem 2 is analogous to skillful driving; it is the kind of thing we might think of when we give up using CAS to mimic paper-and-pencil calculations and accept them instead as new and interesting inhabitants of our mathematical environment.

Do CAS present new ways of fostering mathematical reasoning?

Just as the automobile opened up new worlds of travel to our ancestors, so the CAS opens up new worlds of mathematics. It enables our students to conquer the mathematical Wild West of their imaginations, full of scary critters and dangerous people, by showing them the reality. It is up to us to invite our students to further think about and explore what they see. It is up to us, their teachers, to make sure that their voyages of the mind are like hikes through Yosemite rather than visits to the mall.

Part II

Examples of CAS at Work in the Curriculum and Classroom: Introduction

THE USE of CAS in mathematics classrooms, grades 7–12 and beyond, raises a variety of interesting questions related to the types of examples that can capitalize on the benefits of CAS tools and lead to significant algebra learning. The chapters in this section offer a variety of examples and insight on issues connected with introducing and using CAS. How can we design examples that take advantage of CAS technology? What do these examples look like? Do the examples foster students' interest, motivate students to learn, and help students develop conceptual understanding? Which tasks are inappropriate for CAS use and under what circumstances? What mathematics can now be studied that was not feasible in a paper-and-pencil environment? How should CAS examples be structured to encourage students to engage in "conjecturing" and testing their conjectures?

David Bowers uses a CAS as a medium in which students try to figure out, using pattern matching, the rules for elementary algebraic calculations, before they work through more standard derivations for these rules.

Al Cuoco and Ken Levasseur develop examples of topics and investigations in classical mathematics that become tractable when students use the computational power of CAS.

Todd Edwards presents his own classroom experiences with beginning algebra students using CAS as a tool for exploring expression simplification and for aiding in the acquisition of conceptually supported equation-solving skills.

Tim Garry provides several examples of how a CAS environment can be used in advanced high school courses both as a medium for generating examples to test conjectures and as a tool in constructing proofs of these conjectures.

Jeremy Kahan and Terrence Wyberg give yet another way that CAS environments can be used to introduce important mathematics into the secondary school curriculum: the use of generating functions in combinatorial investigations.

James Schultz offers several examples of appropriate and inappropriate uses of CAS. He also includes an extended example that illustrates how to use symbolic calculations in a CAS to enhance an investigation calculating monthly payments on loans.

Nurit Zehavi and Giora Mann describe the design considerations underlying the construction of two examples drawn from an eighth-grade unit on equations. They also summarize their workshop discussions with teachers who used the two activities in their classes.

The chapters in this section describe just a few of the curricular possibilities that CAS environments make feasible. The goal is to add some specificity to the discussion surrounding their role in secondary-level mathematics courses.

—Carolyn Kieran and Al Cuoco

CHAPTER 5

Promoting Pure Mathematics through Preliminary Investigational Activities Using Computer Algebra

David Bowers

MATHEMATICS is special. It is the only subject in which the teacher provides answers to questions that the students would never have asked themselves.

Behind this bon mot hides a real challenge for us as mathematics educators, namely, how to motivate students and instill a sense of shared ownership of the mathematics that they are expected to learn. This problem seems to be less evident in other curriculum areas—one can imagine young people inquisitive enough to want to know why volcanoes erupt, when Shakespeare lived, or how to reassemble a gear box, but students simply do not of their own volition approach us demanding to know how to rationalize the denominator of a fraction, complete the square on a quadratic, or simplify the sum of two logarithms.

One possible explanation may be that traditional "pure mathematics" classes tend to be one-directional in structure, with the teacher stating what is to be covered and the students simply following the instruction and learning the method. Thus, students arrive for class in a "passive" frame of mind. The only time they become "active" is when they are answering sets of questions that the teacher has deemed necessary, a process that can better be described as "reactive" rather than "proactive."

As a result, students often perceive mathematics as a subject to be accepted on blind faith rather than as a subject that requires full understanding or sound intuition. This perception is encouraged by what I call the "method marks culture," referring to the common practice in England of giving students partial credit for identifying and attempting to apply an appropriate method of solution, even if the subsequent work is riddled with errors and the final answer is hopelessly wrong. Two things disturb me in such a scenario: (1) the implication that creditworthy mathematics is essentially algorithm-driven, and (2) the tolerance that is frequently shown for "silly" answers. This situation is, in my view, a consequence of students' not understanding what they are doing nor why they are doing it. They are simply following a set of abstract rules.

To adapt a well-known saying, good mathematics means never handing in a silly answer. How can we encourage this outcome? We need to give students the chance to develop an intuitive feeling for their work, and doing so means giving them the opportunity to investigate mathematics in more than one learning environment. We need to encourage a sense of ownership of the mathematics on the part of the students by allowing them to make discoveries rather than presenting them with everything "on a plate." The use of computer algebra systems (CAS) can facilitate these aims. The examples of investigational activities presented in this chapter illustrate one possible approach.

DESCRIPTION OF THE ACTIVITIES

An "investigational activity" is defined here as a structured sequence of tasks that works toward a specific goal, allowing students to reach that goal independently and claim the resulting knowledge or insight for themselves. If the investigational activities are completed before the teacher's formal presentation, either in a computer-lab session or in directed private study, then the mathematical results and methods should already have been internalized by the students to some extent. The students should then understand these results and the motivation behind them more readily than if the results had been presented "cold" by the teacher.

The investigational activities proposed here are nothing more than sets of carefully structured questions for which CAS can provide the answers, followed by a space for the students to note the patterns and trends that they have observed and state the underlying rules. By way of illustration I have taken topics from GCE A-level mathematics, a preuniversity course for United Kingdom students ages 16–18.

Example 1—Elementary Factorization

Algebraic factorization is an important skill for higher level mathematics, yet it often represents the first real stumbling block for students as their mathematical studies move from the specific realm of numbers to the general realm of algebra. In my experience, students need time to grasp the processes involved in algebraic factorization, a requirement that we as educators are often guilty of overlooking because the mathematical concept being explored is "obvious" to us.

A CAS-supported preliminary investigational activity would allow the students time to observe the effect of factorization on simple expressions and reflect on the result of the process, away from the pressures of the traditional classroom. The structure of such an activity is outlined below:

- Give the students a list of expressions, such as those below, and ask them to use the CAS to factor the expressions and complete the right-hand side.

$$ab + ac = \underline{\qquad}$$
$$xw + xy - xz = \underline{\qquad}$$
$$pqr + qrs = \underline{\qquad}$$
$$4pw - 6npr = \underline{\qquad}$$

- Ask the students to explain what they observe.

- Give the students another list of expressions in the same format as those listed in item 1 above. Ask them to complete the right-hand side without the use of technology and then use the CAS to check.

- Ask the students to create original two-term or three-term algebraic expressions that they think will factor, predict the results, and then use the CAS to check.

- Ask the students to create original expressions that they think cannot be factored, explain why not, and then use the CAS to check.

Of course, this investigational activity could be developed in a number of ways, for example, by making the link between factoring and expanding more explicit or by including simple powers. The important thing is that students are given the opportunity to work through a short, clearly defined, and purposeful activity that gives them some introductory insight and prepares them for the class lesson in which the teacher will present the topic in a more formal manner. As a result, the students should "hit the ground running."

Example 2—Rules of Logarithms

How do we teach students the essential result shown below?

$$\log(a) + \log(b) = \log(ab), \text{ where } a, b > 0$$

One way is to present them immediately with the "classical" derivation below:

Let $10^p = a$ and $10^q = b$.

Hence $p = \log(a)$ and $q = \log(b)$.

Then $(10^p)(10^q) = 10^{p+q} = ab$.

Hence $p + q = \log(ab)$.

Therefore $\log(a) + \log(b) = \log(ab)$.

In my experience, this derivation bewilders most students, who in the United Kingdom will generally be ages 16–17 when asked to learn the rules of logarithms, having had little previous experience with formal proofs.

An alternative approach is for the teacher simply to state, "Log(a) plus log(b) equals log(ab). Fact! Learn it!" This tactic might appeal to the diligent student who sees a relatively simple piece of rote learning as a way to gain easy marks, but it does not impart much in the way of insight, understanding or ownership of the mathematics. Verifying this rule numerically using the ten-digit display of a standard pocket calculator is a rather messy and unsatisfying experience. Fortunately, CAS will apply the rules of logarithms exactly without reverting to decimal approximation. Thus, for example, log(2)+log(7) simplifies on screen to log(14). This feature is the basis of a preliminary investigational activity that follows the structure outlined below:

- Give the students a list of sums of logarithms, such as those below, and ask them to use the CAS to simplify the expression and complete the right-hand side.

$$\log(2) + \log(3) = \log(\quad)$$
$$\log(5) + \log(7) = \log(\quad)$$
$$\log(2) + \log(13) = \log(\quad)$$

- Ask the students to consider the results and suggest a rule for simplifying the sum of two logarithms.

- Give the students another list of examples in the same format as those listed in item 1 above. Ask them to use their own rule to complete the right-hand side, use the CAS to check, and rethink the rule if discrepancies arise.

- To establish a "both ways" understanding, ask the students to use their rule to complete the left-hand side of a list of examples, such as those that follow:

$$\log(\ ) + \log(\ ) = \log(15)$$
$$\log(\ ) + \log(\ ) = \log(34)$$

- Ask the students to use the CAS to simplify the solutions on the left-hand side and check that the given right-hand side is obtained.

- Ask the students are to state their rule and justify their level of confidence. Does the rule always work? Can they find any examples in which the rule does not work? Can they extend the rule to related cases?

In the light of experience, the numbers included in this investigational activity need to be chosen carefully, lest the CAS be "too clever" and obscure the expected result. For example, the computer algebra system DERIVE will simplify $\log(4) + \log(54)$ as $3\log(6)$. Although this result might pave the way for an extension activity for more able students, an advisable approach is to keep the a and b as prime numbers initially. Also worth noting is the fact that, in general, CAS will not algebraically simplify $\log(a) + \log(b)$ as $\log(ab)$ unless a and b have been declared in advance to be positive.

OBSERVATIONS ON THE USE OF PRELIMINARY INVESTIGATIONAL ACTIVITIES WITH STUDENTS

The philosophy behind these investigational activities is to allow students to gain an intuitive feeling for certain topics and claim a basic understanding of the topics before the formal presentation by the teacher. This method can result in a more satisfying lesson than if the teacher had used the this-is-the-method-now-learn-it approach. After participation in such activities, the students should be familiar with the outcomes of the lesson so they can apply their experience of the activity to subsequent discussion and development of the examples by the teacher.

CAS have opened up the possibilities here by allowing the user to experiment with, and receive feedback in, such areas as algebra and calculus, which hitherto have often only been passively accepted. Furthermore, CAS are patient providers of answers. Students can take as long as they wish to work out what is going on and can enter as many different examples as they need to convince themselves of the validity or invalidity of any conjectures.

Of course, any data-driven pattern recognition that might be generated through such activities does not constitute a proof. I would stress the preliminary nature of these investigational activities. They do not replace teacher-led classroom sessions but supplement and prepare students for the lessons that will be used to present formal justification and proof, at whatever level the teacher deems appropriate. Having already used CAS to investigate some examples, students have the advantage of being more familiar with the outcomes and, hence, are more receptive to a detailed explanation and proof.

I still recall observing a lesson during which the class encountered differential calculus for the very first time. The teacher immediately embarked on an "x-plus-delta-x" proof of the derivative of x^2. After many chalkboards of work, the teacher finally arrived, with a flourish, at $dy/dx = 2x$. The students who had been dutifully copying everything, were, however, totally bemused and failed to see any significance in this last line. If before this lesson, they had been guided to use CAS to investigate what happens when they "differentiate" x^2, x^3, x^4, and so on, they might have already internalized the basic $(n)(x^{n-1})$ result. I am convinced that they would then have followed the proof with greater interest, insight, and understanding. The reader will correctly conclude that I am very much in favor of a use-then-prove approach to college mathematics over the more traditional prove-then-use method.

The use of preliminary investigational activities, such as those outlined above, requires a clear scheme of work through the mathematics syllabus so that the activities can be issued to students just before the topic is formally covered in class.

The availability of an adequate number of computers for the students to use is essential. If the mathematics class has regularly scheduled lessons in a computer lab, then perhaps the last thirty minutes of one of those lessons would be a suitable time to work on an investigational activity. Alternatively, students need to have a clear understanding of what is expected of them if they use computers outside class time in a central college resource area or at home.

The students should already be familiar with the computer algebra system that is used in class so that technical issues do not hold them back. A helpful tactic is to print on the reverse side of the activity sheet an "aide memoir" that summarizes the necessary CAS commands.

An advisable approach is to limit the investigational activity to one sheet of paper and to use the format of a "gapped handout," that is, a computer-lab worksheet with blanks, so that students can write their observations on the handout. They can then easily find and refer to these observations in the subsequent lesson in which the topic is formally covered by the teacher.

The whole point of these preliminary investigational activities is to motivate the students; therefore, all students must be able to achieve at least part of the task, so that they can all make some contribution to the class discussion. For this reason, the activities start with some fairly mundane "button pressing" to obtain from the CAS some routine results for investigation. Even if the less speedy or less determined students get no further than this point, they will have done some relevant preparation for the lesson.

My personal satisfaction from developing and working with this kind of preliminary investigational activity comes from observing the "aha!" effect when a student gradually identifies a rule and tries it out with other examples using the CAS. This result is particularly pleasing if it concerns a topic that I previously struggled to teach to students while I was standing at the chalkboard.

Activity 3

Find the point of intersection of the lines given below. Show your work.

$$3x + 5y = 17$$
$$8x - 3y = -8$$

Solution:

- $3 \cdot (3 \cdot x + 5 \cdot y = 17) + 5 \cdot (8 \cdot x - 3 \cdot y = {}^-8)$

$$49 \cdot x = 11$$

- $\dfrac{49 \cdot x = 11}{49}$ $x = 11/49$

- $8 \cdot (3 \cdot x + 5 \cdot y = 17) - 3 \cdot (8 \cdot x - 3 \cdot y = {}^-8)$

$$49 \cdot y = 160$$

- $\dfrac{49 \cdot y = 160}{49}$ $y = \dfrac{160}{49}$

The point of intersection is $\left(\dfrac{11}{49}, \dfrac{160}{49}\right)$.

CHAPTER 6

Classical Mathematics in the Age of CAS

Al Cuoco
Ken Levasseur

WHEN NUMERICAL calculators first made their way into schools, they were heralded with both enthusiasm and worry. The enthusiasts saw the machines as an avenue for students to move past computational drudgery; they envisioned calculators as an opportunity for teachers to eliminate the pages of identical calculations whose only point was to build efficiency, speed, and accuracy with pointless hand calculations of ever-increasing complexity. The worriers feared a further erosion of students' abilities to calculate with, and reason about, numbers and arithmetic; they warned of a growing population of high school graduates who, without their calculators, would not be able to read the newspaper, multiply a number by 10, or make change for a dollar.

To some extent, both the enthusiasts and the worriers were correct. In the twenty-five years since calculators became a staple of school mathematics, teachers have learned a great deal about how to use them and, just as important, how not to use them.

Introduction

Two decades ago, eighth-grade students nationwide were working the same pages of multiplication problems that eighth graders had always worked, except they were using calculators to find the answers. This scenario is rarely seen today. Instead, while teachers still often use calculators to bypass arithmetic algorithms, teachers and their students also often go "unplugged" and study computational algorithms as mechanisms for understanding the structural properties of the number systems of arithmetic. When students need to find a decimal approximation for 3/7, using a calculator is the most efficient method. When they want to investigate the pattern in the repeating digits for the decimal expansions of *k*/7, calculators are perfect for generating examples. Yet calculators are not as much help in analyzing long division and understanding what students see in the examples. Calculators are wonderful machines for generating data that can lead to conjectures. They are not much help when establishing them.

The grandiose claims of twenty years ago have gradually evolved. Most teachers and educators view numerical calculators as simply one more tool in students' mathematical toolkit. Debates about instruction in calculator use have matured to the point that instead of being about *whether* to use the machine, they are about *how* to use it best.

Numerical calculators, used in school mathematics, have evolved from a simple replacement for paper-and-pencil calculations to a flexible and general-purpose tool with well-defined limitations. Their use has been "mirrored," with minor modifications, in uses of other technology. Graphing software, geometry software, and spreadsheets have all followed the same cycle—starting with heated debates among those educators claiming that traditional "skills" were obsolete and those calling for a total ban on the technology, then settling on a more-or-less general acceptance of the value added by their smart use.

We are about to go through the all the debates again, this time with CAS environments. Although the systems have been making inroads into the undergraduate curriculum for some time, they are only beginning to be taken seriously in precollege education—primarily because of their implementation on calculators. Enthusiasts are calling for a drastic cutback in instruction in paper-and-pencil algebraic calculation, "mindless symbol manipulation," as it has come to be known. Worriers are warning of upcoming college freshmen who will fail calculus because they cannot simplify

$$\frac{1}{x} + \frac{x}{2}$$

without CAS. And in many second-year-algebra classes, students are working the same pages of factoring problems that they have always worked, except they are using a CAS to get the answers.

By writing this chapter, we hope to move the CAS debate forward. We have used CAS technology in both high school and undergraduate classes for several years and have experienced ways to use the systems that enhance student learning, as described in Hibbard and Levasseur (1999) and in *Mathematical Methods* (EDC 2001). We have also come to the conclusion (Cuoco and Manes 2001) that the technology is no replacement for basic mathematical and algebraic understanding. We are beginning to find ways that CAS environments can be used to help students develop such understanding. In the following sections we use examples to illustrate the following points:

1. CAS technology can enhance and make tractable many beautiful classical topics that were previously considered too technical for high school students. This use of technology reduces computational overhead, allowing students to easily perform algebraic transformations and to construct visualizations that would have been impossible or overly distracting without the technology.

2. CAS technology can be used to experiment with algebraic expressions in the same way that calculators can be used to experiment with numbers: generating data, making patterns apparent, and giving students the raw data from which they can generate conjectures. Because formal expressions are "first-class" objects in a CAS, the technology has the potential to bring a renewed and modern emphasis on formal algebra in school mathematics, described by Goldenberg (chapter 1, this volume) as the *algebra of forms*.

3. CAS technology allows students to build models of algebraic objects that have no faithful physical counterparts. Cuoco and Goldenberg (1996) and Cuoco (1997) discuss these uses of technology, which support the point of view that building a computational model for a mathematical structure helps students build the mental constructions they need to "interiorize" that structure. Furthermore, computational models like those provided by Hibbard and Levasseur (1999), Cuoco and Goldenberg (1996), and Cuoco (1997) are *executable*. Therefore, students can turn thought experiments into actual experiments, building a laboratory in which they can conduct investigations, explore conjectures, and look for patterns.

4. Other computational environments allow students to construct models built up from basic objects, such as sets, lists, geometric objects, and functions. CAS environments add the facility to model mathematical objects and to perform generic calculations with algebraic *expressions*: polynomials, rational functions, and formal power series. Kahn and Wyberg (chapter 9, this volume) show how this use of algebraic expressions can introduce students to basic notions about gen-

erating functions. Other applications (see, e.g., Cuoco [1997]) include developing the notion of algebraic isomorphism through systems that "calculate the same." Because formal algebraic expressions are "universal" objects in algebra, CAS environments allow for very general investigations. For example, the rule for multiplying complex numbers, $(a + bi)(c + di) = (ac - bd) + (ad + bc)i$, can be derived by asking a CAS for the remainder when $(a + bx)(c + dx)$ is divided by $x^2 + 1$.

5. Although this book is about CAS environments in their pure form as algebraic calculators, most CAS implementations occur in a much wider computational environment, allowing for visualization, high-powered numerical calculations, and functional programming. These features greatly enhance the usefulness of CAS systems in school mathematics.

Each of the examples in this chapter's next sections has been used in high school classes, undergraduate courses, and professional development seminars for teachers. We do not present the following examples as pieces of classroom-ready curriculum, but such examples can be found in publications by Hibbard and Levasseur (1999), EDC (2001), and EDC (2002a). Instead, they are descriptions of some classical mathematical topics that have the potential, with the use of CAS technology, to come alive for students in grades 10 through 16. We invite you to grab your paper and pencil, pull out a second-year-algebra reference book, and fire up your CAS. We hope you enjoy each example and adapt them for your own classes. Detailed Mathematica code and formatting routines are available for downloading at http://www.edc.org/CME/showcase/. Throughout this chapter, we use the CAS language built into the TI-89 or the TI-92 calculators. Although the latter is a more limited system than Mathematica, it is more widely available and is powerful enough for the basic calculations in the following sections.

Fitting Polynomials to Data, Newton, and Calculations

We begin by using a CAS as an algebraic calculator; more precisely, we use it as a machine that can simplify expressions. Even at this level, we see the potential for helping students learn new mathematics by using the technology.

Many curricula ask students to find a "rule" that can be used to generate a table. For example, suppose we have the following table for a mystery function f:

n	$f(n)$
0	1
1	−1
2	11
3	49
4	125
5	251
6	439
7	701

Of course, infinitely many functions f will "do the trick," even if we restrict our investigation to polynomial functions (see, e.g., EDC [2002a]), but a *unique* polynomial function of minimal degree that agrees with the table does exist. Many methods have been developed over the years to find this function, and most CAS environments implement one or more of these methods as *primitives*.

One such classical method, dating back to Newton, is included in publications by Boole (1860) and EDC (2002a) and connects nicely with a technique used throughout middle and high school, namely, asking students to look at the differences in the table below.

n	$f(n)$	∅
0	1	−2
1	−1	12
2	11	38
3	49	76
4	125	126
5	251	188
6	439	262
7	701	

If you can find a pattern in the Ø column, you may find a formula for f. If a pattern is not apparent, continue "differencing":

n	$f(n)$	Ø	Ø²	Ø³
0	1	−2	14	12
1	−1	12	26	12
2	11	38	38	12
3	49	76	50	12
4	125	126	62	12
5	251	188	74	
6	439	262		
7	701			

You have discovered that the third differences are constant. But you need not stop there. A CAS can generate the complete difference table for any data set. The Mathematica code for doing this procedure can be found at http://www.edc.org/CME/showcase/.

n	$f(n)$	Ø	Ø²	Ø³	Ø⁴	Ø⁵	Ø⁶	Ø⁷
0	1	−2	14	12	0	0	0	0
1	−1	12	26	12	0	0	0	
2	11	38	38	12	0	0		
3	49	76	50	12	0			
4	125	126	62	12				
5	251	188	74					
6	439	262						
7	701							

Notice that, from the way the table is constructed, any element is the sum of its "up and over."

Classical Mathematics in the Age of CAS

n	f(n)	∅	∅	∅
0	1	−2	14	12
1	−1	12	26	12
2	11	38	38	12
3	49	76	50	12
4	125	126	62	12
5	251	188	74	
6	439	262		
7	701			

Therefore, we can take any entry in the $f(n)$ column, replace it by its up-and-over, replace these two numbers by *their* up-and-overs, and keep moving up the table toward the first row.

Next try this strategy for $f(3)$:

$f(3) = 49$

$= 11 + 38$

$= (−1 + 12) + (12 + 26) = −1 + 2 \cdot 12 + 26$

$= (1 + −2) + 2 \cdot (−2 + 14) + 14 + 12 = 1 + 3 \cdot (−2) + 3 \cdot 14 + 12$

Look below at a different version of $f(3)$, this time with some emphasis added and details suppressed:

$f(3) = \mathbf{1} \cdot 49$

$= \mathbf{1} \cdot 11 + \mathbf{1} \cdot 38$

$= \mathbf{1} \cdot −1 + \mathbf{2} \cdot 12 + \mathbf{1} \cdot 26$

$= \mathbf{1} \cdot 1 + \mathbf{3} \cdot (−2) + \mathbf{3} \cdot 14 + \mathbf{1} \cdot 12$

Could it be? Is Pascal's triangle rearing its head?

Try this strategy for $f(4)$:

$f(4) = \mathbf{1} \cdot 125$
$= \mathbf{1} \cdot 49 + \mathbf{1} \cdot 76$
$= \mathbf{1} \cdot (11 + 38) + \mathbf{1} \cdot (38 + 38)$
$= \mathbf{1} \cdot 11 + \mathbf{2} \cdot 38 + \mathbf{1} \cdot 38$
$= \mathbf{1} \cdot (-1 + 12) + \mathbf{2} \cdot (12 + 26) + \mathbf{1} \cdot (26 + 12)$
$= \mathbf{1} \cdot -1 + \mathbf{3} \cdot 12 + \mathbf{3} \cdot 26 + \mathbf{1} \cdot 12$
$= \mathbf{1} \cdot (1 + -2) + \mathbf{3} \cdot (-2 + 14) + \mathbf{3} \cdot (14 + 12) + \mathbf{1} \cdot (12 + 0)$
$= \mathbf{1} \cdot 1 + \mathbf{4} \cdot (-2) + \mathbf{6} \cdot 14 + \mathbf{4} \cdot 12 + \mathbf{1} \cdot 0$

We seem to have discovered a general method for calculating $f(n)$. Can we turn this method into an explicit polynomial formula for f? We can ignore the Ø columns for $e > 3$ because they are all 0. We now have the following:

$f(3) = 49 = \mathbf{1} \cdot 1 + \mathbf{3} \cdot (-2) + \mathbf{3} \cdot 14 + \mathbf{1} \cdot 12$
$f(4) = 125 = \mathbf{1} \cdot 1 + \mathbf{4} \cdot (-2) + \mathbf{6} \cdot 14 + \mathbf{4} \cdot 12$
$f(5) = 251 = \mathbf{1} \cdot 1 + \mathbf{5} \cdot (-2) + \mathbf{10} \cdot 14 + \mathbf{10} \cdot 12$
$f(6) = 439 = \mathbf{1} \cdot 1 + \mathbf{6} \cdot (-2) + \mathbf{15} \cdot 14 + \mathbf{20} \cdot 12$
$f(7) = 701 = \mathbf{1} \cdot 1 + \mathbf{7} \cdot (-2) + \mathbf{21} \cdot 14 + \mathbf{35} \cdot 12$

If we "snake back through" the difference table, winding our way to the top row by replacing each number by its up-and-over, we get a combination of the elements in the first row, and the coefficients are precisely the entries in the "input" row of Pascal's triangle. It appears that we can calculate $f(n)$ by the rule

$$f(n) = 1 \cdot \binom{n}{0} + (-2) \cdot \binom{n}{1} + 14 \cdot \binom{n}{2} + 12 \cdot \binom{n}{3}.$$

This rule certainly works if n is 3, 4, 5, 6, or 7. Since we are looking for a polynomial, we will get rid of the $\binom{n}{k}$ expressions and replace them with their algebraic equivalents:

$$f(n) = 1 \cdot 1 + (-2) \cdot n + 14 \cdot \frac{n(n-1)}{2} + 12 \cdot \frac{n(n-1)(n-2)}{6}$$

A little algebra, by hand or with a CAS, can turn the equation above into a nice cubic:

$$f(n) = 2n^3 + n^2 - 5n + 1$$

We can check that this function agrees with the table.

Most CAS environments have primitives that allow the user to calculate binomial coefficients. On the TI-89, the function is denoted by nCr. This function is typically used to calculate numerical values (so that $_5C_2$ returns $\binom{5}{2}$, or 10). Because expressions are first-class objects in a CAS, we can use the nCr with a *generic* or indeterminate first input to get the form of the binomial coefficient:

$$_nC_r(x, 3)$$

returns

$$\frac{x(x-1)(x-2)}{6},$$

and

$$_nC_r(x, 5)$$

returns

$$\frac{x(x-1)(x-2(x-3)(x-4))}{120}.$$

Here are a few expanded forms of these combinatorial polynomials, generated with a CAS:

k	$\binom{x}{k}$
0	1
1	x
2	$\dfrac{-x + x^2}{2}$
3	$\dfrac{2x - 3x^2 + x^3}{6}$
4	$\dfrac{-6x + 11x^2 - 6x^3 + x^4}{24}$
5	$\dfrac{24x - 50x^2 + 35x^3 - 10x^4 + x^5}{120}$
6	$\dfrac{-120x + 274x^2 - 225x^3 + 85x^4 - 15x^5 + x^6}{720}$
7	$\dfrac{720x - 1764x^2 + 1624x^3 - 735x^4 + 175x^5 - 21x^6 + x^7}{5040}$

We can use the conjectured form of the mystery function f to get the simplified version in one "swoop":

$$1 * {}_nC_r(x,0) + (-2) * {}_nC_r(x,1) + 14 * {}_nC_r(x,2) + 12 * {}_nC_r(x,3)$$

produces

$$2x^3 + x^2 - 5x + 1.$$

This is just the beginning. Newton's difference formula states that our method will always work (see, e.g., EDC [2002a]). Suppose we have a data table whose table has mth differences that are constant and not zero:

Input	Output	Δ	Δ^2	Δ^3	...	Δ^m
0	a_0	a_1	a_2	a_3	...	a_m
1					...	a_m
2						a_m
3						a_m
4						a_m
5						a_m
6						a_m
7						a_m

A polynomial function of degree m that agrees with the table is

$$f(x) = \sum_{k=0}^{m} a_k \binom{x}{k}.$$

Using CAS technology, students can easily find a polynomial that agrees with a table. Suppose we have a table like the one that follows:

Input	Output
0	−5
1	−6
2	7
3	166
4	843
5	2770

The successive differences do not look constant:

Input	Output	Δ	Δ^2	Δ^3	Δ^4	Δ^5
0	−5	−1	14	132	240	120
1	−6	13	146	372	360	
2	7	159	518	732		
3	166	677	1250			
4	843	1927				
5	2770					

Since Δ, Δ^2, Δ^3, Δ^4 are not constant, it can be shown (EDC 2002a) that no polynomial of degrees 1, 2, 3, or 4 will fit the table, but we can find a degree 5 polynomial that works. Furthermore, we know that

$$-5\binom{x}{0} - 1\binom{x}{1} + 14\binom{x}{2} + 132\binom{x}{3} + 240\binom{x}{4} + 120\binom{x}{5}$$

will work, so we whip out our CAS and simplify the polynomial above to obtain the following expression:

$$x^5 - 3x^3 + x^2 - 5$$

Therefore, we have found a fifth-degree polynomial that agrees with the table.

Trigonometric Identities, Chebyshev, and Patterns

In the previous section, we used a CAS to expand and simplify algebraic expressions. In this section, we look at its capabilities to factor expressions and to perform more complicated algebraic transformations.

Trigonometric identities receive less attention in modern trigonometry courses than they did in the past. One reason for this decline in focus is that traditional treatments seemed to depend on a bag of tricks. Some students got the "knack" of establishing the identities, but many other students perceived that proving an identity required just the right ad hoc, out-of-the-blue substitutions and transformations. Many of the exercises seemed like esoteric curiosities; only a few identities seemed to have any application, and those were encountered in calculus courses. A good housecleaning seemed to be in order.

What remains is a set of truly important facts that have numerous applications throughout mathematics and science, including the following:

- *The Pythagorean identity*, which is equivalent to the fact that the unit circle is invariant under reflection in any diameter:

$$\sin^2\theta + \cos^2\theta = 1$$

- *The addition formulas for sine and cosine*, which are equivalent to the fact that the unit circle is invariant under rotation about the origin:

$$\sin(\alpha+\beta) = \cos\alpha\sin\beta + \sin\alpha\cos\beta$$
$$\cos(\alpha+\beta) = \cos\alpha\cos\beta - \sin\alpha\cos\beta$$

For simple proofs that use the algebra of complex numbers, see EDC (2001).

These basic identities imply (by replacing α and β with θ) the *double angle formulas*:

$$\sin(2\theta) = 2\cos\theta\sin\theta$$
$$\cos(2\theta) = \cos 2\theta - \sin 2\theta$$

In the second identity, replace $\sin^2\theta$ by $1-\cos^2\theta$, and we can express as a $\cos(2\theta)$ polynomial in $\cos\theta$:

$$\cos(2\theta) = 2\cos^2\theta - 1$$

What about $\cos(3\theta)$? We can try to proceed inductively, expanding $\cos(3\theta)$ as $\cos(\theta = 2\theta)$:

$$\begin{aligned}\cos(3\theta) &= \cos(\theta+2\theta) \\ &= \cos\theta\cos(2\theta) - \sin\theta\sin(2\theta) \\ &= \cos\theta(2\cos^2\theta - 1) - \sin\theta(2\sin\theta\cos\theta) \\ &= 2\cos^3\theta - \cos\theta - 2\sin^2\theta\cos\theta \\ &= 2\cos^3\theta - \cos\theta - 2(1-\cos^2\theta)\cos\theta \\ &= 4\cos^3\theta - 3\cos\theta\end{aligned}$$

So $\cos(3\theta)$ is a polynomial (a cubic, in fact) in $\cos\theta$. Before we invest in more hand calculation, let us use the CAS to generate some data. The question at hand is "Can $\cos(n\theta)$ be expressed as a polynomial (maybe of degree n) in $\cos\theta$?"

Most CAS implementations have functions that tell the algebraic manipulation routines that trigonometric functions should be treated as rational functions of sine and cosine. For example, Mathematica has functions like **TrigExpand** and **TrigSimplify**; the TI machines have **tExpand** and **tSimplify**. Here is how they work:

$$\text{tExpand }(\cos(3\theta))$$

returns

$$\cos(\theta) - 4(\sin(\theta))^2 \cdot \cos(\theta)$$

This result is not exactly what we obtained by hand but can be turned into it by replacing $\sin^2 \theta$ by $1 - \cos^2 \theta$. Next try it for $\cos(4\theta)$. Typing

$$\text{tExpand }(\cos(4\theta))$$

produces

$$1 - 8(\sin(\theta))^2 \cdot (\cos(\theta))^2.$$

Even before we simplify it, we can see that the answer will be a polynomial of degree 4 in $\cos\theta$. It looks like we have a conjecture in the wings. After simplifying the expression above, we have the following:

$\cos 1\theta$	$\cos\theta$
$\cos 2\theta$	$-1 + 2(\cos\theta)^2$
$\cos 3\theta$	$-3(\cos\theta) + 4(\cos\theta)^3$
$\cos 4\theta$	$1 - 8(\cos\theta)^2 + 8(\cos\theta)^4$

We care most about the *form* of the alleged polynomials. What are their degrees? What can we say about the coefficients? For questions like this, "$\cos\theta$" is just a distraction; it might as well be an x or a y or any other symbol. The CAS can help strip away the distractions for us with its substitution routines. The syntax varies from system to system, but the idea is that the user can replace $\cos\theta$ by any letter desired. On the TI system, typing

$$3a + 2 \,|\, a = 4$$

produces 14. And typing

$$1 - 8(\sin(\theta))^2 \cdot (\cos(\theta))^2 \,|\, \cos(\theta) = x$$

Classical Mathematics in the Age of CAS

produces

$$1 - 8(\sin(\theta))^2 \cdot x^2.$$

The annoying sine term still exists. If $\cos\theta = x$, then $\sin^2\theta = 1 - x^2$:

$$1 - 8(\sin(\theta))^2 \cdot x^2 \,\big|\, (\sin(\theta))^2 = 1 - x^2$$

produces

$$8x^4 - 8x^2 + 1.$$

This double substitution becomes quite mechanical, so it can be automated. We can generate many multiple-angle-generating polynomials, such as those below:

1	x
2	$-1 + 2x^2$
3	$-3x + 4x^3$
4	$1 - 8x^2 + 8x^4$
5	$5x - 20x^3 + 16x^5$
6	$-1 + 18x^2 - 48x^4 + 32x^6$
7	$-7x + 56x^3 - 112x^5 + 64x^7$
8	$1 - 32x^2 + 160x^4 - 256x^6 + 128x^8$
9	$9x - 120x^3 + 432x^5 - 576x^7 + 256x^9$
10	$-1 + 50x^2 - 400x^4 + 1120x^6 - 1280x^8 + 512x^{10}$
11	$-11x + 220x^3 - 1232x^5 + 2816x^7 - 2816x^9 + 1024x^{11}$
12	$1 - 72x^2 + 840x^4 - 3584x^6 + 6912x^8 - 6144x^{10} + 2048x^{12}$
13	$13x - 364x^3 + 2912x^5 - 9984x^7 + 16640x^9 - 13312x^{11} + 4096x^{13}$
14	$-1 + 98x^2 - 1568x^4 + 9408x^6 - 26880x^8 + 39424x^{10} - 28672x^{12} + 8192x^{14}$
15	$-15x + 560x^3 - 6048x^5 + 28800x^7 - 70400x^9 + 92160x^{11} - 61440x^{13} + 16384x^{15}$
16	$1 - 128x^2 + 2688x^4 - 21504x^6 + 84480x^8 - 180224x^{10} + 212992x^{12} - 131072x^{14} + 32768x^{16}$

The nth row yields a formula for $\cos(n\theta)$ (by replacing x by $\cos\theta$), producing as many identities as we like—and many more than our students would care to memorize.

It appears there is a polynomial $T_n(x)$ of degree n that has the property that

$$T_n(\cos\theta) = \cos(n\theta),$$

and we have used a CAS to generate the first seventeen polynomials in the preceding table. Even with a CAS, our method becomes cumbersome for large n. We can look among the T_n for *algebraic* relations that might help reduce the tedium. In particular, if we look for a recursive relation to generate the table, we see that to get the degree right, we have to multiply $T_{n-1}(x)$ by $2x$ to get close to $T_n(x)$. But $2xT_{n-1}(x) \uparrow T_n(x)$; we need to subtract something. A little experimenting shows that what we subtract seems to be $T_{n-2}(x)$. If that is true, as it is for all the polynomials in the table, we have a Fibonacci-like recursive definition to generate the T_n:

$$T_n(x) = \begin{cases} 1 & \text{if } n = 0 \\ x & \text{if } n = 1 \\ 2xT_{n-1}(a) - T_{n-2}(x) & \text{if } n > 2 \end{cases}$$

Using this relation to generate the polynomials in the table is certainly more efficient, and we conjecture that if we extend the table using this rule, we get polynomials that yield multiple-angle formulas. The reader might try proving this conjecture, by induction, for example. For other approaches, see EDC (2002a), Rivlin (1974), and EDC (2002b). Most CAS implementations allow for user-defined functions. On the TI system, we can define a function t so that $t(n) = T_n(x)$:

```
t(n)
Func
If n=0 Then
Return 1
ElseIf n=1 Then
Return x
Else
Return 2*x*t(n−1)−t(n−2)
EndIf
Endfunc
```

These polynomials T_n show up all over mathematics and science—so much so that many CAS environments have then built in as primitives. They are the celebrated *Chebyshev polynomials*, and they have many wonderful algebraic properties. Just as an example, look at their factored forms (which can be generated with a few keystrokes in your CAS):

Factored Chebyshev Polynomials

n	$T_n(x)$
0	1
1	x
2	$-1 + 2x^2$
3	$x(-3 + 4x^2)$
4	$1 - 8x^2 + 8x^4$
5	$x(5 - 20x^2 + 16x^4)$
6	$(-1 + 2x^2)(1 - 16x^2 + 16x^4)$
7	$x(-7 + 56x^2 - 112x^4 + 64x^6)$
8	$1 - 32x^2 + 160x^4 - 256x^6 + 128x^8$
9	$x(3 - 4x^2)(3 - 36x^2 + 96x^4 - 64x^6)$
10	$(-1 + 2x^2)(1 - 48x^2 + 304x^4 - 512x^6 + 256x^8)$
11	$x(-11 + 220x^2 - 1232x^4 + 2816x^6 - 2816x^8 + 1024x^{10})$
12	$(1 - 8x^2 + 8x^4)(1 - 64x^2 + 320x^4 - 512x^6 + 256x^8)$
13	$x(13 - 364x^2 + 2912x^4 - 9984x^6 + 16640x^8 - 13312x^{10} + 4096x^{12})$
14	$(-1 + 2x^2)(1 - 96x^2 + 1376x^4 - 6656x^6 + 13568x^8 - 12288x^{10} + 4096x^{12})$
15	$x(3 - 4x^2)(-5 + 20x^2 - 16x^4)(1 - 32x^2 + 224x^4 - 448x^6 + 256x^8)$
16	$1 - 128x^2 + 2688x^4 - 21504x^6 + 84480x^8 - 180224x^{10} + 212992x^{12} - 131072x^{14} + 32768x^{16}$

At this point we have a laboratory in which we can experiment with all kinds of questions. We might wonder what we can say about m and n if T_m is a factor of T_n. When is T_n irreducible? What can we say about the zeros of the $T_n(x)$? About the graphs of the functions defined by the T_n? Some such graphs, generated in Mathematica, are shown in figures 6.1 and 6.2. A CAS lets us play with these and the many other questions we can find in these data.

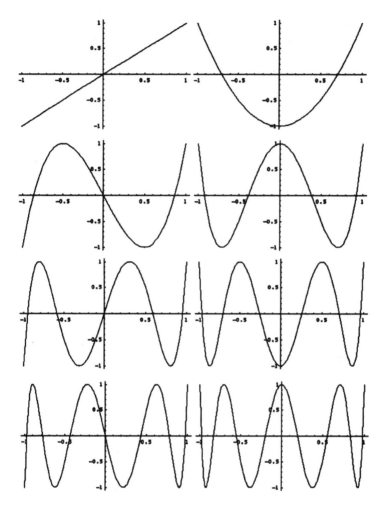

Fig. 6.1. A gallery: $y = T_k(x)$, $1 \leq k \leq 8$

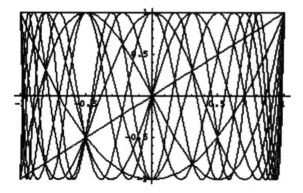

Fig. 6.2. Overlay of the graphs of $y = T_k(x)$, $0 \leq k \leq 10$

And what about proofs? They are left for the reader's enjoyment, along with the following hint: First establish that $T_n \cos\theta = \cos(n\theta)$.

Patterns in Primes, Goldbach, and Computer Mathematics Systems

In addition to the algebraic power of a CAS, the sheer computational power of a typical CAS gives it the potential to enhance non-algebraic segments of the curriculum. Most CAS implementations contain sophisticated graphical routines, mathematical functions that can handle very large inputs with accuracy, and even the capability of evaluating limits. In this sense, rather than CAS, the systems are really computer mathematics systems (CMS). Our final example shows how the power of a CMS can be used to see and analyze something that would be difficult to detect without technology. The idea for this investigation came from Josh Abrams. For this example, we need the power of a CAS implemented on a computer. We used Mathematica here, but Maple, MATLAB, or any of the other available CMS environments would work just as well. Complete Mathematica code for everything in this section is available at http://www.edc.org/CME/showcase/.

Goldbach's conjecture says that every even integer greater than 2 is the sum of two primes. The conjecture, first stated in 1742, remains unanswered despite the attempts of many mathematicians to settle it. If we ask our CAS to plot the number of ways n can be written as a sum of two primes against n, things get interesting for large n (see fig. 6.3). Definite spikes appear, and a little numerical experimentation shows that they seem to be at multiples of 6. The reader might try to explain or establish this behavior. Our point here is that the sheer computational power of some CAS environments can be used to gain insights and to make such conjectures.

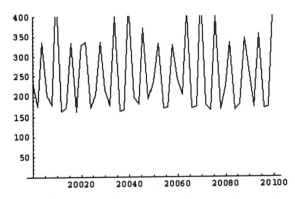

Fig. 6.3. *The number of ways n can be written as the sum of two primes*

Conclusion

The topic of this part of the book is "examples." Although we think each of the previous sections provides concrete examples of productive CAS use, we are not lobbying for including these particular topics in the secondary school or undergraduate curriculum. We certainly do see CAS environments paving the way for a role for such classical topics, involving substantial algebraic calculations, formal algebra, and high-powered scientific computation. And, just as important, we see our examples as exemplars of a *style of work*, a way of investigating, modeling, conjecturing, and explaining, that can be made more accessible to a wider audience in the age of CAS.

REFERENCES

Boole, George. *A Treatise on the Calculus of Finite Differences*. London: Macmillan, 1860.

Cuoco, Al. "Constructing the Complex Numbers." *International Journal of Computers for Mathematics Learning* 2 (1997): 155–86.

Cuoco, Al, and Paul Goldenberg. "A Role for Technology in Mathematics Education." *The Boston University Journal of Education* 178, no. 2 (1996): 15–32.

Cuoco, Al, and Michelle Manes. "When Memory Fails." *Mathematics Teacher* 94(6) (September 2001): 489–93.

EDC. *Mathematical Methods*. Armonk, N.Y.: It's About Time, 2001.

———. *The Mathematical Companion*. Armonk, N.Y.: It's About Time, 2002a.

———. "Making Mathematics: A Website Devoted to Research Projects in Mathematics for Secondary Students and Teachers." [2002b]. Available at www2.edc.org/makingmath. World Wide Web.

Harel, Idit, and Seymour Papert. *Constructionism*. Norwood, N.J.: Ablex Publishing Corporation, 1991.

Hibbard, Allen, and Kenneth Levasseur. *Exploring Abstract Algebra with Mathematica*. New York: Springer-Telos, 1999.

Rivlin, Theodore J. *The Chebychev Polynomials*. New York: John Wiley & Sons, 1974.

CHAPTER 7

Calculator-Based Computer Algebra Systems: Tools for Meaningful Algebraic Understanding

Michael Todd Edwards

AS MINIATURIZATION of calculator-based technologies continues to narrow the gap between "calculator" and "computer," powerful computer algebra systems (CAS) now appear in devices no larger than traditional graphing calculators. Unlike conventional calculators, CAS-equipped devices have the ability to manipulate symbolic expressions. As **figure 7.1** suggests, CAS-equipped calculators enable students to perform a variety of manipulation-oriented tasks—including factoring and simplifying algebraic expressions—without pencil or paper. Algebraic tasks that previously consumed hours of instructional time can now be accomplished with minimal symbolic know-how using a calculator equipped with CAS.

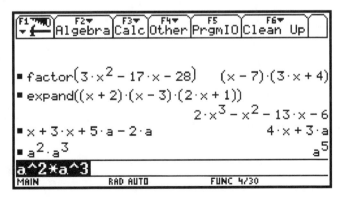

Fig. 7.1. Sample calculations possible with the TI-92 calculator

Introduction

Hand-held CAS, such as the TI-92, quickly and accurately carry out many of the tasks that high school mathematics students have heretofore performed with slower, more error-prone pencil-and-paper techniques. The existence of CAS-equipped tools brings into question the continued emphasis on by-hand manipulative procedures at the secondary school level. Heid (1995, p. 143) notes the following:

> The technological world in which students and teachers now operate demands a radical transfiguration of algebra in schools. As we enter the twenty-first century, the study of algebra in schools must focus on helping students describe and explain the world around them rather than on developing and refining their execution of by-hand symbolic-manipulation procedures—procedures that are better accomplished through the informed use of computing tools.

The use of CAS promises to shift the emphasis of algebra instruction away from technical aspects of symbolic manipulation and toward pattern recognition and conceptual understanding of calculator-generated output.

To anticipate the impact that CAS will have on teaching and learning symbolic manipulation, it is useful to examine the impact that handheld graphers had on *graphing* instruction a decade ago. Handheld graphers changed the way that students and teachers approached graphing in school classrooms:

1. With the graphers, students were able to produce graphs more quickly and more accurately.

2. The study of families of graphs became possible. Thus, the transformational qualities of graphs became more central to instruction.

3. The emphasis shifted away from the by-hand production of graphs, and increased attention was placed on interpretation of graphical data.

The impact of CAS on secondary school mathematics instruction parallels the changes brought about by graphing calculators a generation earlier:

1. Students are able to produce simplified algebraic expressions and solutions *more quickly* and *more accurately*.

2. The study of families of solutions and families of simplifications is possible. Thus, *pattern recognition* becomes a focus of algebra instruction.

3. The emphasis shifts away from the by-hand production of algebraic forms, and increased attention is placed on *interpretation of algebraic data*.

In this chapter, the use of the TI-92 as a learning tool to enhance students' understanding of symbolic procedures and algebraic structures is discussed. Two fundamental methods for using CAS with secondary school students are proposed: (1) as a tool for exploring expression simplification, and (2) as a tool for aiding in the acquisition of equation-solving skills. The examples provided serve to illustrate the ways in which the development of conceptual understanding can be fostered while using CAS in mathematical areas that have traditionally been the exclusive domain of paper-and-pencil symbol manipulation.

Exploring Expression Simplification

As a secondary school mathematics teacher, I have experienced success using the TI-92 calculator with novice algebra students. The ability of the TI-92 to generate numerous solutions quickly and accurately enables students to build conjectures regarding algebraic patterns that they see on the calculator's screen. Drijvers (1995, p. 5) made the following observation:

> This exploration phase can lead to interesting discoveries and forms the basis of the explanation phases. In the later phase the results of the explorations will be sorted and proved, or they will lead to the development of new concepts.

As I illustrate below, the calculator not only encourages students to form hypotheses, it also proves useful as a motivator for classroom discussion. Furthermore, CAS-based investigations encourage teachers to reexamine student misconceptions regarding algebraic symbolism. At the same time, CAS-based investigations provide instructors with possible teaching remedies for these misconceptions.

Expressions with nonnegative terms and one variable

Whenever I introduce concepts to students, I find the projection capabilities of the TI-92 particularly helpful. For instance, when first discussing the notion of "combining like terms," I display a series of warm-up problems from my calculator onto a projection screen. As **figure 7.2** suggests, my discussion typically begins with simple algebraic examples.

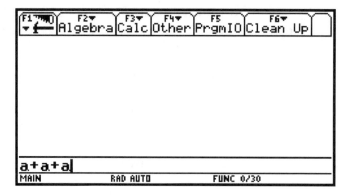

Fig. 7.2. Expressions are initially typed into the input editor of the TI-92.

Before pressing my calculator's ENTER key, I engage my students in a discussion of possible calculator outputs.

> Teacher: (*Pointing to calculator screen*) How many times have I added "*a*" to itself?
>
> Student: (*Waving hand furiously*) Three!
>
> Teacher: (*Pausing, scratching chin thoughtfully*) Hmm! Interesting! (*To entire class*) Three times I've added "*a*." Based on your classmate's observation, write down what you think the calculator will print to the screen once I press ENTER.

As I walk around the room, I ask students to jot their predictions on scrap paper. With few exceptions, students claim that "*three times a*" will appear on the calculator's home screen. This outcome is shown in **figure 7.3**.

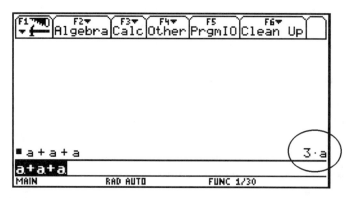

Fig. 7.3. Calculation results appear on the right-hand side of the TI-92 screen.

Typically, my classes investigate several examples of repeated addition involving one variable. Lengthier examples, such as the one shown in **fig-**

ure 7.4, help students grasp the convenience and necessity of the multiplication operator.

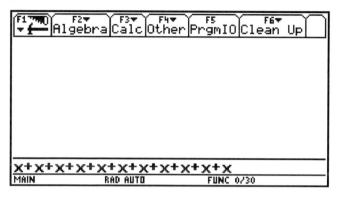

Fig. 7.4. Repeated addition becomes burdensome as more terms are added.

With the expression "$x + x + x + x + x + x + x + x + x + x + x$" displayed on an overhead calculator, I may pose a variety of questions to my students.

- How will the calculator simplify "$x + x + x + x + x + x + x + x + x + x + x$"?
- Why are "$x + x + x + x + x + x + x + x + x + x + x$" and the calculator's answer equivalent?
- In several sentences, explain a possible motivation for using multiplication instead of repeated addition. Refer to our last example in your explanation.

Student responses such as the following are typical:

> The calculator will say $11 \cdot x$. This is because you're adding x together 11 times. When you're adding over and over again, it's easier to just write down the number of times x is added. That is what multiplication shows us here.

While using the TI-92 to simplify expressions with coefficients, students are given opportunities for conceptual growth. For instance, repeated-addition examples encourage students to consider why such expressions as "$2x + 3x + 7x$" and "$12x$" are equivalent. As shown in **figure 7.5**, parentheses may be written directly on the calculator screen to highlight the theory underlying coefficient addition.

To encourage students to thoughtfully examine the patterns they see on their calculators, teachers may find it helpful to assign short writing exercises

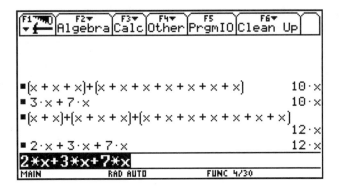

Fig. 7.5. Adding parentheses to the calculator view screen helps students better understand simplification.

dealing with CAS-generated output. Carefully designed questions require students to explain basic algebraic ideas suggested by CAS. Sample questions are provided below.

- Simplify each of the expressions on the left with your TI-92. Write down the simplified form of each expression in the space provided at the right. The first expression has been simplified for you.

Unsimplified Expression	Simplified Expression
$4x + 9x$	$13x$
$2x + 10x$	_____
$9a + 4a$	_____
$12m + 3m + 6m$	_____

- For each item above, a relationship exists between the coefficients of the unsimplified expression and the coefficient of the simplified expression. In several sentences, describe this relationship. Include examples of unsimplified and simplified expressions in your response.

- Typing "$3x + 7x$" into the TI-92 results in the expression "$10x$" being printed to the calculator's screen. In several sentences, justify the correctness of this calculation with a sound logical argument. Include the idea of repeated addition in your response.

As the previous exercises suggest, the use of the TI-92 as a tool for testing conjectures helps teachers focus students' attention on the conceptual underpinnings of algebraic calculations. The following example involving negative terms further illustrates this notion.

Expressions with negative terms and one variable

Contrasting "negative" and "subtract" with the TI-92. After investigating multiplication as repeated addition, a natural approach is to turn students' attention to repeated subtraction. **Figure 7.6** illustrates a typical repeated-subtraction example.

$$-a \quad -a \quad -a \quad -a$$

Fig. 7.6. *Chalkboard example illustrating* $^{-}4a$

When asked to predict how the calculator will simplify the expression in **figure 7.6**, novice algebra students often give a number of conflicting responses. For instance, some students predict that the calculator will output "$4 - a$."

> Since we are subtracting four times, you write down "Four times subtracting a." When you write this out in symbols, you get "$4 - a$."

Other students indicate that the calculator will output "$4 \cdot a$." Two negatives make a positive, so the first two a's make "$2 \cdot a$." Another makes "$3 \cdot a$," so the last one makes "$4 \cdot a$." Another group of students claims that the calculator will generate "$-4 \cdot a$." Typically, the TI-92 helps resolve student conflict involving symbolic calculations. However, when students type the expression "$-a-a-a-a$" by using the subtract key repeatedly, they realize that the TI-92 misinterprets the intent of the user in this instance. The subtraction operation requires a minuend. Hence, when students type "−" using the subtraction key, the function "ans(1)" appears prior to the first term, as shown in **figure 7.7**.

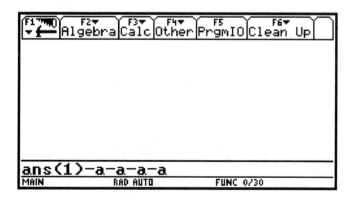

Fig. 7.7. *The subtraction symbol requires two operands on the TI-92.*

When "ans(1) $-a-a-a-a$" is evaluated by the calculator, a variety of unsatisfactory solutions are possible, including cryptic error messages. Students' failure to distinguish between "subtraction" and "negative" is largely responsible for the unexpected calculator behavior. Because the TI-92 keyboard includes *separate* subtraction and negative keys, the calculator provides a meaningful context for exploring distinctions between "subtract" and "negative." As I have noted below in a passage from my personal teaching journal, the key setup of the TI-92 has influenced the way I teach algebra.

> My first encounter using CAS to simplify expressions with negative leading coefficients came in front of a room of frustrated teenagers. I had often glossed over differences between the symbols in my own teaching. For instance, I commonly used the term "minus" to indicate both subtraction and negativity. In fact, prior to work with the TI-92, I might have referred to "5 – 6" as "five minus six" and "-7" as "minus seven." It's safe to assume that other math teachers do the same. That's why I spend so much time discussing "subtraction" and "negative" keys on the calculator.

Exploration with arithmetic examples aids students' understanding of the "subtraction" and "negative" operators. For instance, the teacher may first ask students to predict the result of entering "5 – 2" (i.e., 5 subtract 2) into the calculator. The teacher may then ask students to predict how the calculator will simplify "5 -2" (i.e., 5 negative 2). As illustrated in **figure 7.8**, the latter calculation produces an unexpected result. Careful examination of calculator output provides the impetus for further classroom discussion. As students consider examples involving subtraction *and* negative signs, they begin to realize that subtraction requires *two* operands, whereas the negative symbol is *unary*, requiring only *one* operand.

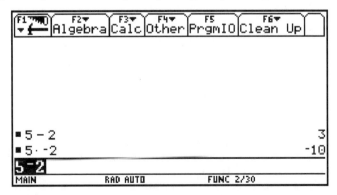

Fig. 7.8. Students are initially surprised when they discover that the inputs "5 – 2" and "5 -2" generate different outputs.

Equipped with a stronger understanding of both "negative" and "subtraction," students can use the TI-92 to explore basic examples involving addition and subtraction of like terms. In particular, the equivalence of "adding a negative" and "subtracting," another stumbling block for novice students, can be emphasized with CAS-based examples. **Figure 7.9** illustrates that expressions such as "$7x - 3x$" and "$7x + {}^-3x$" are equivalent.

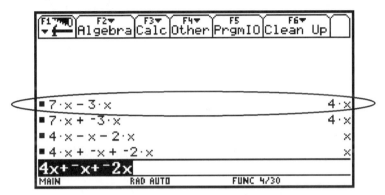

Fig. 7.9. Symbolic examples emphasize the equivalence between "subtraction" and "adding a negative."

A natural extension of the equivalence exploration above involves simplifying expressions containing variables and constants. Because the TI-92 has a tendency to simplify expressions immediately, skipping intermediate steps of expression simplification, I sometimes use the calculator as a "conversation piece" *after* students have simplified algebraic expressions for several class meetings *without* the TI-92. Students may enter an expression such as "$-7x + 5 - 3x - 8$" into the TI-92. As **figure 7.10** illustrates, the calculator generates the simplified expression "$-10x - 3$" as output.

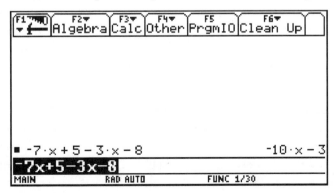

Fig. 7.10. A more complicated problem simplified with the TI-92

Referring to the expression $-7x + 5 - 3x - 8$ on the calculator, the teacher may ask students the following questions:

- Why did entering "$-7x + 5 - 3x - 8$" into the calculator generate the term "$-10 \cdot x$" in the answer?
- What terms from the input were combined to form "$-10 \cdot x$"?
- Why did entering "$-7x + 5 - 3x - 8$" into the calculator generate the term "-3" in the answer?
- What terms from the input were combined to form "-3"?
- A classmate tells you that the expressions "$-7x + 5 - 3x - 8$," "$-10x - 3$," and "$-10x + -3$" are really the same. Do you agree or disagree with this statement? Explain your response.

Development of Equation-Solving Skills

Previous examples in this chapter have highlighted the TI-92's automatic simplification of expressions. Although automatic simplification may be useful for students as they look for algebraic patterns, many students want to see intermediate algebraic steps, particularly when they are learning how to solve equations. Students have good reasons for wanting to see intermediate steps. After all, step-by-step equation solving allows them to review important concepts, such as equivalency, inverse, domain, and function. Many of my students claim they have a better understanding of algebra after solving equations in a step-by-step fashion.

In the article "Should Students Learn Integration Rules," Buchberger (1990) describes two distinct phases through which students pass when learning mathematics, a "black box" phase and a "white box" phase. Buchberger describes the phases in the following manner.

> In the "white box" phase, algorithms must be studied thoroughly, i.e. the underlying theory must be treated completely and algorithmic examples must be studied in all details. In the black box phase, problem instances from the area can be solved using symbolic computation software systems [with intermediate steps hidden from the user] (p. 10).

In the following paragraphs, we explore ways that the TI-92 can be used during the white-box phase as students learn to solve equations. Calculator examples highlight the ability of CAS to solve equations in a step-by-step manner. Throughout this section, calculator techniques that highlight individual algebraic steps are referred to as white-box methods. Likewise, techniques that hide individual steps are referred to as black-box methods.

Solving Linear Equations

In the discussion below, the equation "$2x + 9 = {}^-x + 3$" is solved using the TI-92 in a white-box fashion. The techniques shown are adapted from Kutzler (1996.) Students first type the equation into the TI-92 and press ENTER. The result of entering the expression is shown on the bottom right of the screen, as illustrated in **figure 7.11**.

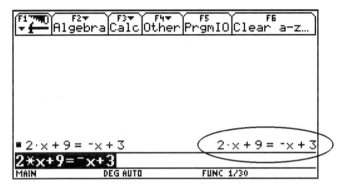

Fig. 7.11. Entering an equation to solve from the TI-92 input editor

Using binary operators, such as addition, subtraction, multiplication, and division, students apply various algebraic transformations to each side of the equation "$2x + 9 = {}^-x + 3$" in an attempt to isolate the variable x. As **figure 7.12** illustrates, when 9 is subtracted from the equation "$2x + 9 = {}^-x + 3$," the calculator displays "$2x = {}^-x - 6$" as output.

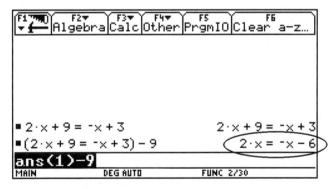

Fig. 7.12. The TI-92 displays results of intermediate calculations along the right-hand side of the calculator's home screen.

Students can next add x to each side of the equation "$2x = {}^-x - 6$." This step is illustrated in **figure 7.13**.

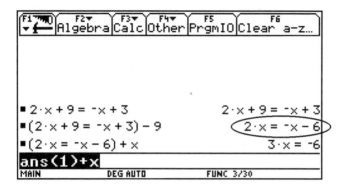

Fig. 7.13. Further intermediate calculations on the TI-92

When solving the resulting equation "$3x = {-}6$" for x, many students mistakenly subtract 3 from each side to "get rid of the 3." Unfortunately, the results of such an operation do not "cancel out" the coefficient "3," as shown in **figure 7.14**.

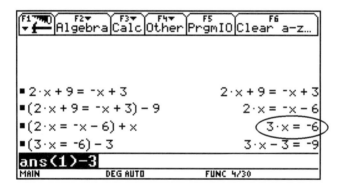

Fig. 7.14. With the TI-92, students are able to apply productive and nonproductive transformations in an attempt to solve equations.

Using the delete key with the TI-92, students can "clear off" incorrect steps and try again. Eventually students realize that division by 3 cancels out multiplication by 3. The equation solution appears in the bottom-right-hand corner of the home screen, as illustrated in **figure 7.15**.

Using the calculator in a "white-box" fashion to solve equations offers students certain benefits that traditional pencil-and-paper methods do not. For instance, because the calculator correctly applies whatever transforma-

Calculator-Based Computer Algebra Systems

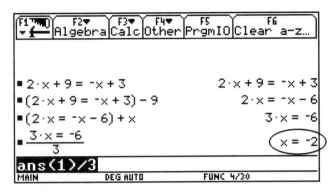

Fig. 7.15. *Solving a linear equation successfully using the TI-92*

tion the student selects, students are free to focus attention on *appropriate transformations*—without the fear of performing an arithmetic procedure incorrectly. Some students use the calculator in lieu of pencil and paper; other students perform calculations by hand first and then replicate their steps on the calculator to intercept errors. One of my students noted, "I . . . like how the calculator shows all of the steps, so if you made 'handmade' mistakes, you could find where you went wrong." CAS give students opportunities to consider equation solving from a conceptual perspective. This idea is explored more thoroughly in the following paragraphs.

Checking Solutions to Equations

Although many students learn correct *procedures* for solving linear equations, their mastery of such tasks does not guarantee that they *conceptually* grasp equation solving. The ability of the TI-92 to represent solutions from multiple perspectives—symbolic, tabular, and graphical—encourages students to consider equation solving from more conceptual points of view.

Checking answers from the home screen

After solving an equation, students can employ symbolic methods from the TI-92 home screen to check their solutions. For instance, suppose students wish to verify that "-2" is a correct solution to the equation "$2x + 9 = -x + 3$." The TI-92 *with* (|) operator provides students with a method for substituting specific values for x. As **figure 7.16** illustrates, the input "$2x + 9 = -x + 3 | x = -2$" substitutes the value of -2 for x.

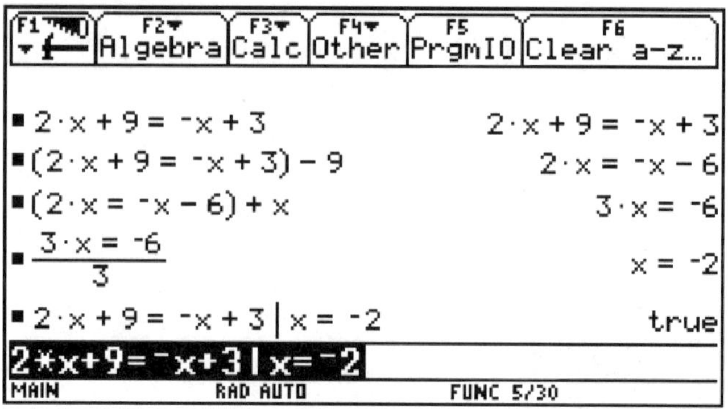

```
┌F1──┐┌─F2─┐┌─F3─┐┌─F4─┐┌──F5──┐┌──F6──┐
│ ▾f │Algebra│Calc│Other│PrgmIO│Clear a-z...│
```

■ 2·x + 9 = ‾x + 3   2·x + 9 = ‾x + 3
■ (2·x + 9 = ‾x + 3) − 9   2·x = ‾x − 6
■ (2·x = ‾x − 6) + x   3·x = ‾6
■ $\frac{3 \cdot x = \text{-}6}{3}$
                                            x = ‾2
■ 2·x + 9 = ‾x + 3 | x = ‾2   true
2*x+9=‾x+3 | x=‾2
MAIN   RAD AUTO   FUNC 5/30

Fig. 7.16. The TI-92 with (|) operator can be used to test the "truthfulness" of various arithmetic sentences.

As students evaluate the equation "$2x + 9 = {-}x + 3$" for various values of x, they find that values other than ‾2 render "$2x + 9 = {-}x + 3$" a false statement, as illustrated in **figure 7.17**. Students are encouraged to discuss why some values of x make equations true statements but others do not.

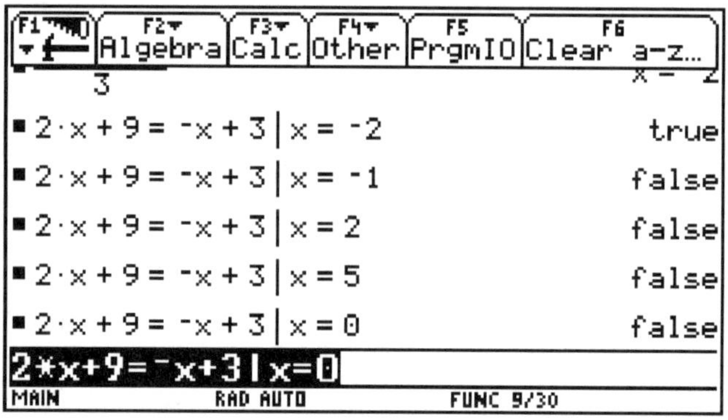

```
┌F1──┐┌─F2─┐┌─F3─┐┌─F4─┐┌──F5──┐┌──F6──┐
│ ▾f │Algebra│Calc│Other│PrgmIO│Clear a-z...│
```
    3   x = ‾2
■ 2·x + 9 = ‾x + 3 | x = ‾2   true
■ 2·x + 9 = ‾x + 3 | x = ‾1   false
■ 2·x + 9 = ‾x + 3 | x = 2   false
■ 2·x + 9 = ‾x + 3 | x = 5   false
■ 2·x + 9 = ‾x + 3 | x = 0   false
2*x+9=‾x+3 | x=0
MAIN   RAD AUTO   FUNC 9/30

Fig. 7.17. Students are encouraged to consider why "$2x + 9 = {-}x + 3$" is true at ‾2.

Checking answers with tables

Alternatively, when checking answers with tables, students can be encouraged to look at *each side* of an equation as a *separate* function. For instance, in the example "$2x + 9 = {-}x + 3$," students can view "$2x + 9$" and "${-}x + 3$" as distinct functions of x. In solving the equation, they will find input values that produce the same output for both functions. This task is readily accomplished with the calculator's table utility. As **figure 7.18** illustrates, the functions "$2x + 9$" and "${-}x + 3$" generate equivalent output when $x = {-}2$.

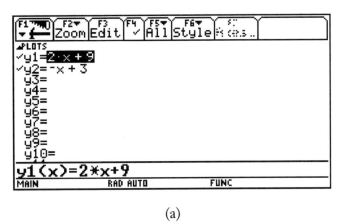

Fig. 7.18. (a) Students first enter each side of the equation "$2x + 9 = {}^-x + 3$" into the equation editor of the TI-92. (b) The TI-92 table utility indicates that the functions return the same value when $x = {}^-2$.

Checking answers with graphs

In the secondary school curriculum, students spend considerable time exploring graphs of linear equations. Viewing simple equations as sets of simultaneous graphs fosters a visual interpretation of equation solving. That students may not automatically make the connection between equations and graphs is interesting to notice. By providing simultaneous graphs of "$2x + 9$" and "${}^-x + 3$," teachers help students link previously understood content, such as graphing lines, with symbolic methods of equation solving.

Students discover that the intersection of the functions "$2x + 9$" and "${}^-x + 3$" and the solution to the equation "$2x + 9 = {}^-x + 3$" are linked; the x-coordinate of the intersection is the same as the solution to the equation. By examining problems graphically, students are encouraged to consider equation solving as a method for calculating intersections of two curves, as illustrated in **figure 7.19**.

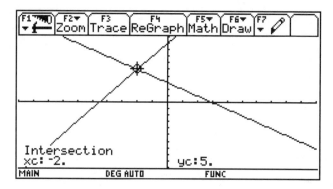

Fig. 7.19. Solving equations by locating intersections of functions on a graphing calculator

The calculator's ability to depict symbolic information in a graphical manner proves a helpful learning aid for many students who are visual learners. One advanced algebra student noted the following:

> The calculator more than anything speeds up the process that my brain normally takes. It helps me get my answers out quicker ... *graphs have also helped me tremendously because I am such a visual learner. It helps me to be able to picture the equation even before I solve it* [emphasis added].

To fully capitalize on the calculator's ability to represent equations as tables, graphs, and symbols, teachers need to ask questions that require students to consider equations from multiple perspectives. When used in conjunction with conceptually based activities, the multirepresentational aspects of CAS may help novice algebra students better grasp the theory of equation solving.

Conclusion

Throughout this chapter, I have illustrated ways that the TI-92 can serve as a worthwhile teaching tool in secondary school algebra classrooms. Whether used for exploration of expression simplification or for skill development in equation solving, the CAS seamlessly links algebraic concepts with symbolic work. When used to project calculator-generated solutions onto an overhead screen, the CAS is particularly helpful in encouraging classroom discourse. Because CAS can readily simplify expressions and solve equations that are found in most secondary school textbooks, the TI-92 permits the more technical aspects of symbolic manipulation to be accompanied by new conceptual emphases.

Nevertheless, the TI-92's ability to automatically simplify expressions needs to be approached with caution. Students often remark that the calculator "does too much" work for them—particularly when they are learning new material. Buchberger (1990) recommends that in their initial learning stages, students require repeated exposure to underlying theory and algorithms to build sufficient conceptual understanding of various algebraic topics. The examples given in this chapter illustrate how this conceptual understanding can be fostered while using CAS. However, to fully capitalize on the calculator's ability to highlight concepts, teachers need to ask questions that require students to consider algebraic ideas from a more conceptual point of view.

Acknowledgments

The author would like to extend many thanks to the students and faculty of Upper Arlington High School, Columbus, Ohio, for encouraging the use of handheld CAS during his five-year tenure at the school. In addition, he would like to sincerely thank the students and faculty of Linworth Alternative School, Worthington, Ohio, for all of their kind support and words of encouragement throughout the writing process. Lastly, he would like to thank Carolyn Kieran for her timely and thoughtful editorial advice and Jim Kozman, of Franklin Heights High School, for his insight regarding pedagogical uses of CAS with secondary school students.

REFERENCES

Buchberger, Bruno. "Should Students Learn Integration Rules?" *Sigsam Bulletin* 24(1) (1 January 1990): 10–17.

Demana, Franklin, and Bert Waits. "The Role of Hand-Held Computer Symbolic Algebra in Mathematics Education in the Twenty-First Century: A Call for Action!" Paper presented at NCTM Standards 2000 Technology Conference, Washington, D.C., June 1998.

Drijvers, Paul. "The Use of Graphics Calculators and Computer Algebra Systems: Differences and Similarities." *International Derive Journal* 1(1) (1994): 71–82.

———. "White-Box/Black-Box Revisited." *International Derive Journal* 2(1) (April 1995): 3–13.

Edwards, Michael Todd. "The Electronic "Other": A Study of Calculator-Based Symbolic Manipulation Utilities with Secondary School Mathematics Students." Unpublished doctoral dissertation, Ohio State University, 2001.

Heid, M. Kathleen. *Algebra in a Technological World*. Reston, Va.: National Council of Teachers of Mathematics, 1995.

Kieran, Carolyn. "The Learning and Teaching of School Algebra." In *Handbook of Research on Mathematics Teaching and Learning*, edited by Douglas Grouws, pp. 390–418. New York: Macmillan, 1992.

Kozman, James, Michael T. Edwards, and Bert Waits. "Using Computer Algebra Calculators to Teach Better Mathematics Better." Paper presented at the Annual Meeting of the National Council of Teachers of Mathematics, Chicago, Illinois, April 2000.

Kutzler, Bernhard. *Improving Mathematics Teaching with DERIVE*. Lancashire, England: Chartwell-Yourke, 1996.

———. *Solving Linear Equations with the TI-92*. Hagenberg, Austria: BK Teachware, 1997.

Texas Instruments. TI-92. Dallas, Tex.: Texas Instruments, 1997. Calculator.

Activity 4

Given the rational expression

$$\frac{x^3 - 9x^2 + 14x}{-x^2 - 3x + 10}$$

(a) Reduce the expression.

(b) For what values of x (if any) is the original expression not defined? Explain how you know.

(c) For what values of x (if any) is the original expression not equal to the reduced expression? Explain how you know.

Solution:

- $\dfrac{x^3 - 9 \cdot x^2 + 14 \cdot x}{^-x^2 - 3 \cdot x + 10}$ $\dfrac{^-x \cdot (x - 7)}{x + 5}$
- factor $(x^3 - 9 \cdot x^2 + 14 \cdot x)$ $x \cdot (x-7) \cdot (x-2)$
- factor $(^-x^2 - 3 \cdot x + 10)$ $^-(x-2) \cdot (x+5)$

Answer:

(a) $\dfrac{-x(x-7)}{x+5}$

(b) Looking at the third line on the screen, we determine that the denominator is zero at $x = 2$ and $x = -5$. Since division by zero is not defined, the original expression will not be defined for these two values.

(c) Since the reduced expression is defined at $x = 2$ and the original expression is not, the two expressions are equal everywhere except $x = 2$. Neither expression is defined at $x = -5$.

Chapter 8

Computing, Conjecturing, and Confirming with a CAS Tool

Tim Garry

ANY OF my students may decide to purchase a hand-held calculator with a computer algebra system (CAS.) As a teacher of secondary school mathematics, I recognize that students have access to user-friendly CAS technology and I must seriously consider the implications of CAS access as they relate to my teaching and to my students' learning. In this chapter, I approach the issues surrounding CAS and the teaching of mathematics from the point of view that a CAS device is essentially another tool that can be added to the "mathematics tool kits" of students and teachers.

A student acquires *concrete* tools—including a compass, a calculator, and a textbook—and develops *intellectual* tools, such as algorithms, theorems, and methods of proof. As a mathematics teacher, I consider it my responsibility to help my students learn not only the specific characteristics of their concrete and intellectual tools but also *how* and *when* to use the tools in an appropriate and effective manner. Most mathematics teachers have probably had the unsettling experience of observing a student use a calculator to compute a simple sum or product. The analogous situation in a calculus class is to observe a student using a CAS tool to compute a simple derivative or integral instead of quickly computing the derivative or integral mentally. Students and teachers are often unaware of practical ways to exploit powerful tools, such as graphing calculators and CAS, to enhance learning and teaching. Any tool in their mathematics tool kits can be neglected or used inappropriately, but I suggest that a CAS tool probably possesses the greatest potential for neglect or misuse. Mathematics teachers need to seriously consider ways of capitalizing on the power, versatility, and ease-of-use of a CAS tool to support and expedite their teaching objectives, even if they decide that the CAS is inappropriate for a particular lesson or course. My decision to incorporate a CAS tool into my secondary classroom curriculum has proved to be enlightening—both for my students and for myself.

In this chapter, I share two exercises for which I have found the use of a CAS tool appropriate, effective, and enriching. I have used both exercises in several of my advanced secondary mathematics courses and wish to explain how I used the exercises and what I observed from my students' efforts in completing them. The two exercises—to be explained in detail later in this chapter—were not specifically designed for CAS use. Therefore, it is possible for students to complete the exercises without CAS tools. In fact, teachers may prefer that students make an initial attempt to work the exercises without access to a CAS tool. However, by adding a CAS device to students' repertoire of tools, teachers may be able to transform exercises that have been "mired" in tedious computations or incorrect results into opportunities for students to concentrate on analyzing, conjecturing, and verifying.

I like mathematical exercises that generate discussion and questions about *when* and *how* to use the various tools with which students are familiar, whether the tools are algorithms or calculators. Students who are aware of the tools in their mathematics tool kits and the strengths and weaknesses of their tools will be more creative, efficient, and successful in mathematics. Although I have chosen the two exercises in this chapter to highlight the enriching effects of integrating a CAS tool into the curriculum, many classroom situations exist for which I do not allow students to use a CAS tool. Learning and practicing manual techniques in mathematics are important ingredients in students' developing their mathematical aptitudes. Whenever students become active participants in carefully performing a detailed—and perhaps quite rigorous—mathematical algorithm, they gain a new perspective on the "inner workings" of mathematics. Students can improve their analytical skills and conceptual understanding by better understanding procedures and algorithms. In my experience, students display greater skill and discretion in using a CAS tool *after* demonstrating competence in certain manual methods.

The stages of mathematical thought through which the following two exercises guide students cover a range of essential skills, an important reason that I personally like the exercises I have chosen for this chapter. The application and appropriateness of the CAS tool can differ significantly during each stage of students' mathematical thought. The chart below gives a very general outline of the stages of development through which students may proceed in the two exercises that follow.

Mathematical Thinking/Work	Application of a CAS Tool
Mechanical/algebraic computation	Quick and precise results
Analyzing/conjecturing	Efficiently display and investigate patterns
Verifying/proving	Accurately check specific examples and generalizations

I wish to reiterate that a CAS tool is not required for either of the exercises in this chapter. I have used these two exercises both when a CAS tool was available to students and when it was not. Occasionally, students who have access to a CAS tool choose not to use it. The differing degrees of success and the richness of the learning experience associated with the two exercises are interesting to contrast in light of whether a student was or was not using a CAS tool.

Exercise 1 (adapted from Larson, Hostetler, and Edwards [1994])

(a) Write down the first three derivatives of xe^{-x}.

(b) Suggest a formula for

$$\frac{d^n}{dx^n}\left(xe^{-x}\right)$$

that is true for all positive integers n.

(c) Prove that your formula is true by using mathematical induction.

My primary objectives for students who work exercise 1 are that they achieve the following: (1) gain practice with standard differentiation rules—product rule, chain rule and the derivative of the exponential function, (2) formulate a general rule from specific results, and (3) complete a proof by mathematical induction. I have found exercise 1 above to be a valuable and constructive opportunity to engage students in a variety of skills, whether a CAS tool is or is not available. Each time I have used the preceding exercise, I have found myself particularly interested in the last two objectives, parts (b) and (c), and in my students' ability to form their own conjectures. In my experience, the use of a CAS tool in exercise 1 makes the shift of emphasis from mechanical to analytical richer and easier for students.

As a result of CAS use, major obstacles are either eliminated or greatly reduced for students who make a mistake in computing the derivatives in part (a) of exercise 1 or who are unable to observe a pattern in the expressions for successive derivatives. Although most students who do not use a CAS tool are able to correctly compute the first three derivatives for part (a) of exercise 1, they are often unable to observe a pattern and hence get "stuck" or make an incorrect conjecture for part (b) of exercise 1 because they may be unable or unwilling to rewrite the computed derivatives in a form that makes the pattern more recognizable. Students often lack significant experience in conjecturing a general result. I require my students to manually perform the work shown below for computing the first three derivatives. Although some students will recognize a pattern or part of the pattern from the results shown below, many will not.

$$\frac{d}{dx}\left(xe^{-x}\right) = \frac{d}{dx}(x) \cdot e^{-x} + x \cdot \frac{d}{dx}\left(e^{-x}\right)$$
$$= e^{-x} - xe^{-x}$$

$$\frac{d^2}{dx^2}\left(xe^{-x}\right) = \frac{d}{dx}\left(e^{-x} - xe^{-x}\right) = \frac{d}{dx}\left(e^{-x}\right) - \frac{d}{dx}\left(xe^{-x}\right)$$
$$= -e^{-x} - \left(\frac{d}{dx}(x) \cdot e^{-x} - x \cdot \frac{d}{dx}\left(e^{-x}\right)\right)$$
$$= -e^{-x} - \left(e^{-x} - xe^{-x}\right)$$
$$= -2e^{-x} + xe^{-x}$$

$$\frac{d^3}{dx^3}\left(xe^{-x}\right) = \frac{d}{dx}\left(-2e^{-x} + xe^{-x}\right)$$
$$= -2\frac{d}{dx}\left(e^{-x}\right) + \frac{d}{dx}\left(xe^{-x}\right)$$
$$= 2e^{-x} + \frac{d}{dx}(x) \cdot e^{-x} + x \cdot \frac{d}{dx}\left(e^{-x}\right)$$
$$= 2e^{-x} + e^{-x} - xe^{-x}$$
$$= 3e^{-x} - xe^{-x}$$

Students who use a CAS tool will typically carry out the steps shown below. Notice that when using the TI-89, the expressions for the derivatives are automatically displayed in factored form.

$$\frac{d}{dx}(xe^{-x}) = (1-x) \cdot e^{-x} \qquad \frac{d}{dx}((1-x) \cdot e^{-x}) = (x-2) \cdot e^{-x}$$

$$\frac{d}{dx}((x-2) \cdot e^{-x}) = (3-x) \cdot e^{-x} \qquad \frac{d}{dx}((3-x) \cdot e^{-x}) = (x-4) \cdot e^{-x}$$

$$\frac{d}{dx}((x-4) \cdot e^{-x}) = (5-x) \cdot e^{-x} \qquad \frac{d}{dx}((5-x) \cdot e^{-x}) = (x-6) \cdot e^{-x}$$

Although a clear pattern emerges above from the results with a CAS tool, some students may have difficulty forming a conjecture for the general formula. My suggestion to these students is that they compute more derivatives, perhaps the next six. My suggestion is usually met with an immediate groan from students who do not have a CAS tool, because of the lengthy computations. However, students who have a CAS tool often seem delighted to compute more derivatives. If more information is available to students, especially if it is acquired efficiently, then students are more likely to "see" a pattern and form a correct conjecture.

A CAS tool can assist not only in computing and conjecturing but also in confirming results. A conjecture ensuing from the manual computation of the first three derivatives and further consecutive derivatives on a CAS tool could be expressed as

$$\frac{d^n}{dx^n}(xe^{-x}) = (-1)^n(x-n)e^{-x}.$$

The conjectured formula is then defined as a function on the CAS tool, for example, **d(x,n)**. Students are encouraged to evaluate the function for some values of n and compare the CAS results with their previous manual work. Such a comparison supplies convincing evidence toward verifying the proposed formula, but students should anticipate that more work is required to satisfy the demands of a mathematical proof.

Define dn(x,n) = $(-1)^n(x-n)e^{-x}$ dn(x,1) = $-(x-1)e^{-x}$

dn(x,2) = $(x-2)e^{-x}$ dn(x,3) = $-(x-3)e^{-x}$

The confirmation process is completed with a formal proof, in this case, proof by induction in part (c) of exercise 1. A CAS tool can assist nicely in this proof, as outlined below.

Prove by mathematical induction that $\frac{d^n}{dx^n}(xe^{-x}) = (-1)^n(x-n)e^{-x}$ *for any positive integer n.*

Proof steps by hand:	Steps on a CAS tool:
(i) Show that the formula is true for $n = 1$. $$\frac{d}{dx}\left(xe^{-x}\right) = e^{-x} - xe^{-x}$$ $$= -(x-1)e^{-x}$$	$$dn(x, 1) = -(x-1)e^{-x}$$
(ii) Assume the formula is true for $n = k$. $$\frac{d^k}{dx^k}\left(xe^{-x}\right) = (-1)^k(x-k)e^{-x}$$	$$dn(x,k) = (-1)^k(x-k)e^{-x}$$
(iii) Prove the formula is true for $n = k + 1$. $$\frac{d^{k+1}}{dx^{k+1}}\left(xe^{-x}\right) = \frac{d}{dx}\left(\frac{d^k}{dx^k}\left(xe^{-x}\right)\right)$$ $$= \frac{d}{dx}(dn(x,k))$$ $$= \frac{d}{dx}\left((-1)^k(x-k)e^{-x}\right)$$ $$= (-1)^k\left(e^{-x} - (x-k)e^{-x}\right)$$ $$= -(-1)^k(x-k-1)e^{-x}$$ $$= dn(x, k+1)$$	$$dn(x, k+1) = -(-1)^k(x-k-1)e^{-x}$$
(In other words, show that the derivative of the 'kth' derivative of xe^{-x} is the same as the formula when $n = k + 1$.)	(CAS tool finds derivative of formula in terms of k(assumption) and then evaluates formula for $n = k + 1$; verify that they are equal.)

I emphasize to students that they are responsible for understanding the logic of the inductive proof and that they should be able to complete such a proof without a CAS tool. Nevertheless, applying the CAS tool, especially in the final step, can be a real boost for some students.

The development of students' analytical and conjecturing skills can be enhanced by using a CAS tool—if the tool efficiently provides accurate information, displays the information so as to facilitate pattern recognition, removes the tedious work of possible lengthy computations, aids in the verification of specific results, and possibly even performs some of the steps of a proof. Students using a CAS tool also tend to exhibit greater confidence in their conjecture and may benefit from other positive spin-offs associated with using a CAS tool.

Computing, Conjecturing, and Confirming with a CAS Tool

In most classes, a few students are naturally intrigued by the versatility and power of a CAS tool. For example, one of my students wrote a TI-89 calculator program that computes the nth derivative of any entered function (see appendix), thereby taking to another level my suggestion to compute more derivatives. Stimulated by the exercise, the student proceeded to investigate general rules for the nth derivative of other functions. I enjoy knowing that a CAS tool can encourage and facilitate my students' natural curiosity and interest in extending a problem.

A CAS tool can also provide a way to use exercise 1 with students when the absence of specific skill development would normally have prevented its use. I usually use exercise 1 after students have learned and manually practiced various standard differentiation rules and have experienced proof by induction. However, teachers who wish to give students practice in forming and proving a formula by induction only may choose to use this exercise before students have had significant manual practice in differentiation techniques—thereby realizing another advantage of students' having access to a CAS tool.

Exercise 2 (adapted from International Baccalaureate Organisation [1993])

Evaluate each integral for $n = 0, 1, 2,$ and 3. Use the results to obtain a general rule for the integral for any positive integer n, and test your result for $n = 4$.

(a) $\int x^n \ln(x)\, dx$ (b) $\int x^n e^x\, dx$

To ensure that a lesson or exercise is instructive and thought-provoking rather than trivial, teachers must develop awareness of those manipulations that a particular CAS tool can, and cannot, perform. The CAS tool typically used by my students enrolled in Advanced Placement calculus is a TI-89. In part (a) of exercise 2 above, the process of analyzing to generate a general rule for the integral of is rendered meaningless by using the TI-89 calculator, because the calculator formulates the rule immediately—provided the CAS tool does not have a specific value stored for the variable n. However, when the TI-89 is asked to determine the general rule for the integral of $x^n e^x$ in part (b) of exercise 2 above, it is unable to do so.

(a) $\int x^n \ln(x)\, dx = \left[\dfrac{x \ln(x)}{n+1} - \dfrac{x}{(n+1)^2} \right] x^n$ (b) $\int x^n e^x\, dx = \int x^n e^x\, dx$

Hence, the general integral in part (a) of exercise 2 becomes trivial for my students and they focus their analytical energy on the integral in part (b) of exercise 2. Before my students shift focus, I like to encourage them to check the CAS-generated formula for exercise 2, part (a) by manually computing the intergral using integration by parts for at least two values of n.

$n = 0$:

$$\int \ln x \, dx = \ln x \cdot x - \int x \cdot \frac{1}{x} dx$$

$$= x \ln(x) - \int dx$$

$$= x \ln(x) - x + C$$

$n = 1$:

$$\int x \ln x \, dx = (\ln x) \frac{x^2}{2} - \int \frac{x^2}{2} \cdot \frac{1}{x} dx$$

$$= \frac{x^2}{2} \ln(x) - \frac{1}{2} \int x \, dx$$

$$= \frac{x^2 \ln x}{2} - \frac{x^2}{4} + C$$

The process of evaluating the formula for different values of n is streamlined by defining the formula as a function (e.g., intn(x,n)) in terms of x and n on the CAS tool.

$$\text{Define intn}(x, n) = \left[\frac{x \ln(x)}{n+1} - \frac{x}{(n+1)^2} \right] x^n$$

$$\text{intn}(x, 0) = x \cdot \ln(x) - x$$

$$\text{intn}(x, 1) = \frac{x^2 \cdot (2 \cdot \ln(x) - 1)}{4}$$

$$\text{expand } \frac{x^2 \cdot (2 \cdot \ln(x) - 1)}{4} = \frac{x^2 \cdot \ln(x)}{2} - \frac{x^2}{4}$$

In contrast with exercise 1, I always use exercise 2 with my students *after* they have become familiar with manually executing the necessary integration techniques. I want them to consider the implication of adding CAS to their tool kits. With the development of affordable handheld CAS tools, students are very tempted to believe that the purchase of such a tool will instantly transform them into mathematical whizzes. Early in the school year, I often sense from many students—in their words or their behavior— that they think their math achievement will definitely improve because they are the proud owners of CAS tools. I always devote class time to exercises such as the two shown in this chapter and try to make a CAS tool available to all my students. Students who enjoy manual rigor or who are uncomfortable with the technology may choose not to use CAS. Some students— never completely serious—even express the idea that using CAS is tantamount to cheating. Therefore, some carefully selected exercises, such as exercise 2, can encourage students to examine some of their preconceived notions about a CAS tool.

For example, many students who used a CAS tool to evaluate the integral in part (b) of exercise 2 for different values of n considered themselves "smarter" than if they had not used a CAS. Certainly they avoided the careless technical errors that some students made without a CAS tool. As the students proceeded to the task of finding a general rule for the integral for any positive integer n, I detected a shift in many students' perceptions. At this point in the exercise, students' reactions ranged from confidently searching the calculator manual or keyboard for a command that would produce the desired general formula—especially if they had this experience in part (a) of exercise 2—to realizing that no established recipe nor CAS feature for developing the general formula exists. Performing integration by parts, either manually or with CAS, was familiar ground for the students. I encouraged students who were uncertain how to search for a pattern not to "give up" on using the CAS tool to help them form a conjecture. If a student was feeling "stuck", I suggested evaluating the integral for values of n greater than 3. Students without a CAS tool reacted unfavorably to this idea. Therefore, I was able to easily convince most students that CAS was the best tool for the job. The students also developed a growing awareness that the primary focus or challenge of the foregoing exercise was to form and test their own conjectures—my primary objective for asking them to complete exercise 2. The students began to view the CAS tool as an instrument that could aid them in analysis and supply them with information quickly rather than do all the work for them; they no longer viewed it as a tool that was not helpful in the conjecturing stage.

A majority of students have considerable difficulty recognizing a pattern after manually evaluating the integral for $n = 0, 1, 2$ and 3, even if they wisely wrote the results in factored form. With a CAS tool, I recommend that students streamline their work by defining the general integral as a function in terms of x and n (e.g., intn(x,n)), an experience they often recall from exercise 1. They then evaluate the integral for several values of n very quickly, and the results are displayed in factored form.

$$\text{Define intn}(x,n) = \int (x^n * e^x, dx)$$
$$\text{intn}(x, 0) = e^x$$
$$\text{intn}(x, 1) = (x - 1) \cdot e^x$$
$$\text{intn}(x, 2) = (x^2 - 2x + 2) \cdot e^x$$
$$\text{intn}(x, 3) = (x^3 - 3x^2 + 6x - 6) \cdot e^x$$
$$\text{intn}(x, 4) = (x^4 - 4x^3 + 12x^2 - 24x + 24) \cdot e^x$$
$$\text{intn}(x, 5) = (x^5 - 5x^4 + 20x^3 - 60x^2 + 120x - 120) \cdot e^x$$
$$\text{intn}(x, 6) = (x^6 - 6x^5 + 30x^4 - 120x^3 + 360x^2 - 720x + 720) \cdot e^x$$

At this point, students are certainly able to observe some aspects of the pattern, such as the alternating sign, the descending exponent, and the value of the first two coefficients. Invariably, a student will mention that the pattern appears similar to the expansion of a binomial or Pascal's triangle. This observation then leads to writing out Pascal's triangle and comparing it to the pattern of the coefficients of the polynomial in each integral after e^x has been factored out, referred to as the "integral" coefficients. Students recall that the values in Pascal's triangle are binomial coefficients, each of which can be computed from the formula for *combinations* of n items chosen k at a time:

$${}_nC_k = \frac{n!}{k!(n-k)!},$$

for k from 0 to n.

		k						
		0	1	2	3	4	5	6
	0	1						
	1	1	1					
n	2	1	2	1				
	3	1	3	3	1			
	4	1	4	6	4	1		
	5	1	5	10	10	5	1	
	6	1	6	15	20	15	6	1

binomial coefficients

		k						
		0	1	2	3	4	5	6
	0	1						
	1	1	1					
n	2	1	2	2				
	3	1	3	6	6			
	4	1	4	12	24	24		
	5	1	5	20	60	120	120	
	6	1	6	30	120	360	720	720

"integral" coefficients

It is quite unusual for a student to realize, without further investigation, that the numbers in the "integral" coefficients pattern are the numbers for *permutations* of n items chosen k at a time:

$$_nP_k = \frac{n!}{(n-k)!}$$

By closely scrutinizing both patterns of numbers shown in the preceding charts, students can draw conclusions and gain insight. The observation that the first two columns ($k = 0, 1$) in both patterns contain identical values can motivate students to look for a relationship in the columns for $k \geq 2$. Students can observe that the values in a column in the "integral"-coefficients pattern are a constant multiple of the corresponding values in the same column in the binomial-coefficients pattern. For $k = 2$ the multiplying constant is 2, for $k = 3$ it is 6, for $k = 4$ it is 24, for $k = 5$ it is 120, and for $k = 6$ it is 720. Students can conclude that the multiplying constant for each column k is $k!$. Therefore, the pattern for the "integral" coefficients can be generated by multiplying each value in Pascal's triangle, that is, each binomial coefficient, by $k!$. Performing this operation on the formula for *combinations* of n items chosen k at a time gives the formula for permutations of n items chosen k at a time, or

$$\frac{n!}{(n-k)!}.$$

Combining this result with the pattern of alternating signs and descending exponents produces the general formula:

$$\int x^n e^x dx = \sum_{k=0}^{n}\left((-1)^k \frac{n!}{(n-k)!} x^{n-k}\right) e^x$$

This conjecture can be readily confirmed on a CAS tool by defining the general formula as a function in terms of x and n (e.g., g(x,n)) and evaluating for several values of n.

$$\text{Define } g(x,n) = \sum\left((-1)^k \cdot \frac{n!}{(n-k)!} \cdot x^{n-k}, k, 0, n\right) \cdot e^x$$

$$= \sum_{k=0}^{n}\left((-1)^k \cdot \frac{n!}{(n-k)!} \cdot x^{n-k}\right) \cdot e^x$$

$$g(x, 0) = e^x$$
$$g(x, 1) = (x - 1) \cdot e^x$$
$$g(x, 2) = (x^2 - 2x + 2) \cdot e^x$$
$$g(x, 3) = (x^3 - 3x^2 + 6x - 6) \cdot e^x$$
$$g(x, 4) = (x^4 - 4x^3 + 12x^2 - 24x + 24) \cdot e^x$$
$$g(x, 5) = (x^5 - 5x^4 + 20x^3 - 60x^2 + 120x - 120) \cdot e^x$$
$$g(x, 6) = (x^6 - 6x^5 + 30x^4 - 120x^3 + 360x^2 - 720x + 720) \cdot e^x$$

Although the results above do not constitute a formal proof, they present convincing evidence that supports the validity of the conjecture. One possible extension of the exploration exercise above is to ask students to formally prove the general formula.

After working on the exercise described above, most students acknowledge that the CAS tool did not make any "insights" nor decide what information to look at and how to look at it. They tend to recognize that such insights occurred as a result of talking with each other, recalling facts and concepts, and talking with the teacher. Nevertheless, students agree that the CAS tool provided essential information in such a way that they did not become overly frustrated. Therefore, they were able to continue the analysis toward a productive end.

In working through exercises 1 and 2 above, with CAS tools in their hands, my students had a more productive and satisfying experience than if they had not used CAS tools. Many parts of an exercise, such as exercise 2, part (a), become simple and trivial with a CAS tool. If a teacher wants students to gain important practice in a manual technique or algorithm—a worthy objective—then the teacher must consider limiting access to a CAS tool. The implications of student access to a CAS tool are many and varied. Teachers will certainly reflect on what mathematics is taught and how it is taught—an important issue that is beyond the scope of this chapter. My intent in writing this chapter is to share my ideas and relate my observations of students as they use a CAS tool to enhance their mathematical thinking. The two exercises presented in this chapter help students gain a proper perspective on the addition of CAS tools to their mathematical tool kits while engaging and challenging them to compute, make conjectures, and confirm.

Computing, Conjecturing, and Confirming with a CAS Tool

APPENDIX

TI-89 Calculator Program

Finding multiple derivatives for a given function $\dfrac{d^n}{dx^n}(f(x))$

```
()
Prgm

"Program for TI-89 that finds a single
"nth derivative or all the nth derivatives
"from 1 to n for a given function
Lbl  f
ClrIO
Disp   "    derivatives of f(x)"
Disp   ""
Disp   ""

"Input function and n
InputStr   "f(x) = ",fn
expr(fn)»fn
fn»gn
Define   d(fn,x)=Func
¶(fn,x)
EndFunc
Lbl  n
PopUp   {"only nth deriv.","show all 1 to
n"},r
ClrIO
Input   "n = ",n

"Find all derivatives from 1 to n and
"store them in a list called derivs
newList(n)»derivs
For   k,1,n
d(fn,x)»fn
fn»derivs[k]
EndFor

"Display single nth derivative
If r=1 Then
Disp   derivs[n]
Pause
Goto  w
EndIf
```

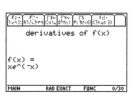

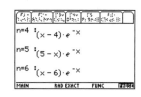

```
"Display all of the nth derivatives
"from 1 to n - 3 at a time on screen
If r=2 Then
ClrIO
0»i
For  k,1,n
Output   25*i,0,"n="
Output   25*i,11,k
Output   25*i,23,":"
Output   25*i+2,28,derivs[k]
Pause
If  k=n
Goto  m
k/3»h
int(h)»g
If h=g Then
ClrIO
a1»i
EndIf
i+1»i
Lbl  m
EndFor
EndIf
Output   15,0,"all stored in list 'derivs'"
Pause

"Menu of choices
Lbl  w
PopUp   {"new n","new function","quit"},q
If q=1 Then
gn»fn
Goto  n
EndIf
If  q=2
Goto  f
If  q=3
Goto  e
Lbl  e
DispHome
EndPrgm
```

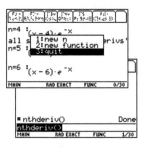

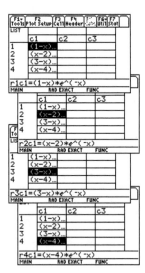

BIBLIOGRAPHY

Larson, Roland E., Robert B. Hostetler, and Bruce E. Edwards. *Calculus*. 5th ed. Lexington, Mass.: D.C. Heath & Co., 1994.

TI-89 Guidebook. Dallas, Tex.: Texas Instruments, 1998.

International Baccalaureate Organisation (IBO). *Mathematics Higher Level Paper 2 Exam, May 1993*. Geneva, Switzerland: IBO, 1993.

CHAPTER 9

Technology Matters: An Invitation to Generating Functions with CAS

Jeremy A. Kahan
Terrence R. Wyberg

IN THE PAST, we were fond of saying, "It's not the technology, it's how you use it." We previously believed that skilled teachers would make good use of technology to teach the content that was important and to promote deeper understanding; other teachers might use technology to drill students in rote, tried-and-tired procedures, or as a substitute for procedural knowledge that remains essential for students.

We no longer say, "It's not the technology, it's how you use it." We now believe that it *is* the technology—at least sometimes. In this chapter, we address—both as teachers and as learners—the issue of using computer algebra systems (CAS) to work with the mathematical topic of generating functions. We do not presume that the reader has prior knowledge of generating functions, because this topic is not presently in the mainstream K–12 curriculum, a situation that may change as a result of the availability of CAS. In this chapter, we explore two problems that introduce the topic of *generating functions* and the power of using CAS to teach this topic.

By exploring the topic of generating functions, we gain a richer perspective on problems we could have solved by other methods, and sometimes we gain the power to do problems that would have been highly impractical to attempt otherwise—problems that go beyond the scope of this chapter. In working with generating functions, the CAS often frees our minds to explore. We could often do without a CAS, but we may choose to use it simply to avoid the hassle of by-hand calculation. Yet sometimes the

CAS is indispensable. Whether using CAS simply for exploring or for solving problems that require their use, exploring generating functions with CAS leads to mathematically significant ideas.

Problem 1: "How Many Games Should You Play?"

The following problem is adapted from *Contemporary Mathematics in Context* (Coxford et al. 1997, p. 513), using suggestions from our student Scott Storla.

> My daughter's team, the Cyclones, has just made the finals of the state hockey tournament, where they will face the Hornets. On the basis of their regular season games, my daughter estimates they have a 60 percent chance of winning any particular game. The tournament organizers are trying to decide whether to stage a one-game, three-game, or five-game series and have asked the teams what they want. My daughter asked me for advice about for what her team should request. What should I tell her?

In solving the How Many Games Should You Play? problem, we use $(c + h)^x$ as a generating function and explore outcomes that represent the number of games each team wins, thus illustrating which team wins the series. Wilf (1994, p. 1) informally defines *generating functions* as "a clothesline on which we hang up a sequence of numbers for display," a metaphor explaining that the coefficients and the exponents of a function provide information that we can use. In the function shown above, we let x vary over the number of games played (one, three, or five) and use the exponents and coefficients of the resulting terms to gain insight into the likelihood of obtaining specific outcomes.

When $x = 1$, we get $1c^1 + 1h^1$. The number 1, acting as the coefficient of c and of h, indicates that each outcome can occur in one way; the exponents track the number of "wins" for each team. We know that the probability of the Cyclones' winning is .6, so if we let c represent the probability that the Cyclones win and let h represent the probability that the Hornets win, we can evaluate the terms and produce the desired result. Does the result hold true for higher values of x? Bernoulli's theorem, dating back to 1713, states that the result will hold true. According to Kline (1972, pp. 273–74), Bernoulli's theorem states that

> ...if p is the probability of a single event happening and q the probability of its failing to happen, then the probability of the event happening at least m times in n trials (where

$m'' n$) is the sum of the terms in the expansion of $(p + q)^n$ from p^n to the term involving $p^m q^{n-m}$.

The mathematics of the year 1713, still relevant and elegant, is more accessible as a result of CAS use. For the three-game series from the problem above, we can choose to use or not use CAS to expand $(c + h)^3$, obtaining $c^3 + 3c^2h + 3ch^2 + h^3$. In the preceding expression, the terms $c^3 + 3c^2h$ represent the ways the Cyclones win the series—a concept that can be verified both directly and by applying Bernoulli's theorem. Next we assign a value of .6 to c and .4 to h and use CAS to evaluate the expression to yield .648. The number .648 represents the probability the Cyclones will win a three-game series, a probability that is greater than the probability of .6 for the team's winning a one-game series.

In the foregoing problem, using the CAS to calculate the probability of the Cyclones' winning a three-game series is not easier than doing the work by hand, especially if we use Pascal's triangle in our by-hand work. However, we recognize advantages of using the CAS to find the probability of the Cyclones' winning in a five-game series. After deleting the values we assigned to c and h, we expand $(c + h)^5$, obtaining $c^5 + 5c^4h + 10c^3h^2 + 10c^2h^3 + 5ch^4 + h^5$. We again assign a value of .6 to c and .4 to h, then edit the expression to $c^5 + 5c^4h + 10c^3h^2$ and evaluate to obtain .68256, the probability of the Cyclones' winning a five-game series. This information is important in making a recommendation to the daughter who plays for the Cyclones. This process can be extended to analyze longer series of games and determine other probabilities.

Several possible methods can be used to solve the tournament problem. *Contemporary Mathematics in Context* approaches the problem by repeatedly simulating it with a random-number table. We can also use such technologies as small computer or calculator programs, dynamic statistics software (e.g., Fathom [Finzer 2001]), and the second box model at http://illuminations.nctm.org/imath/6-8/BoxModel/index.html. One simulation that fosters students' insight involves dropping 4 zeroes and 6 ones into the box, setting the number of draws to the number of games in the series, and exploring the relative frequencies of the outcomes. Such an approach is called a *Monte Carlo method*. Students can also use a tree diagram to approach the problem. The generating function is not the only approach, and readers are encouraged to reflect on the relative merits of the various approaches to the problem.

One weakness in using the generating-function approach to solve this problem is that it does not correspond exactly to what would happen in the real-life scenario described in the problem. If the Cyclones win the first three games, which should happen $(0.6)^3$ of the time, the series is over and no more games are played. On a tree diagram, we can represent this possibility by stopping the "branch" after three games and writing the probability next to it. However, no neat analogous step occurs in the generating function. Instead, we chose to imagine that the last two games were played. Because the outcome is already determined, the probability should not change. The Cyclones still win the series, but the wins are now counted in the c^5 term and parts of the $5c^4h + 10c^3h^2$ terms. After three Cyclone wins, the Cyclones have one way to reach five wins, two ways to reach four wins, and one way to reach three wins for a five-game series, so we need to consider $c^5 + 2c^4h + 1c^3h^2$ for our probability. But

$$c^5 + 2c^4h + 1c^3h^2 = c^3(c^2 + 2ch + h^2) = c^3(c+h)^2 = c^3(1)^2 = c^3,$$

so we get the same result as if we had stopped after three wins. This example illustrates what Tucker (1984, p. 232) sees as a strength of generating functions, that "the algebraic techniques automatically do the combinatorial reasoning for us." We concede that this strength of generating functions can also be a weakness, because we may want to understand the combinatorial reasoning or address its points of difficulty.

In this problem, Dienes (1960) suggests that a teacher should not think about the problem by trying to decide whether the best approach is drawing tree diagrams, using generating functions, or performing a simulation. In exploring the tree diagram approach and the generating function approach, we discover that an underlying structure connects seemingly disparate representations. We call this connecting structure "mathematics," and we want students to find it by reflecting on what the multiple representations have in common. Otherwise, they miss the forest for the trees. Each of the approaches mentioned previously appeals to different students, and because each approach has relative merits, teachers should discuss the strengths and merits of all approaches with their students. Dienes (1960) is in agreement with this recommendation, and *Principles and Standards for School Mathematics* (NCTM 2000, p. 70) states that "it is important for students to reflect on their use of representations to develop an understanding of the relative strengths and weaknesses of various representations for different purposes."

In working with generating functions, students often ask what the variables represent. In some fundamental sense, the variables do not represent anything. In an article classifying uses of variables in algebra, Usiskin (1988) argues that in algebra approached structurally, the variables are arbitrary marks on paper that do not stand for numbers; this approach is the approach taken by CAS, assuming values have not been assigned to the variables. These variables are formal symbols from which polynomials or other algebraic expressions are built. In generating-function work, the variables often serve as placeholders for the exponents and coefficients—the real focus of this chapter. In the tournament problem, the variables play a dual role. On the one hand, they are *variables* in the sense that we took x ranging over the number of games and could consider c and h as taking values from 0 to 1 (while keeping $c + h = 1$) and exploring the changing probabilities while we vary the parameters. On the other hand, we first need to treat c and h as *symbols* so we can perform the polynomial operations on them. If we assign the values too early, we get a constant function of 1 and thus gain no insight. Students usually first encounter this duality in calculus, where we might look for $f'(x)$ when $x = 2$. They first need to take the derivative symbolically, then substitute the value 2 for x. Often, they substitute the value first, decide they are looking at the derivative of a constant, and incorrectly answer 0. This dual sense may also cause difficulty in our example, and we recommend that teachers explicitly discuss the subject with students. We do not shy away from helping students see both ways in which c and h are used; both approaches are important types of mathematical thought.

We acknowledge that a structural approach to algebra may be uncomfortable, at least initially. Teachers usually introduce the commutative property of multiplication using specific values and then ask students to generalize the property using variables as "pattern generalizers," a strategy associated with Usiskin's (1988, p. 11) premise of "algebra as generalized arithmetic." By building on students' knowledge of arithmetic—the "familiar," we are able to motivate them to tackle algebra—the "unfamiliar."

Because students need to think of the variables *formally* when they are working with generating functions, many educators believe that beginning algebra students are not developmentally ready for this topic. Although we agree that students should develop an informal approach to variables and functions before developing a formal approach, we think that the variable, as an arbitrary symbol, can be taught on an informal level *if* students have CAS support. CAS make it easier for students to see how polynomial multiplication does the probability work for them and frees their minds to think about the functions rather than the calculations. Also, the "willingness" of CAS to work with the variable as a pure algebraic symbol without demanding its value or values may help students envision variables in the same way. At the secondary school level, we do not advocate a purely formal algebra program that defines polynomials only as rings involving the symbol x raised to whole powers and coefficients taken from some field, but we do think that the algebraic side of algebra is worthy of students' exploration. By introducing students to informal and example-based structural use of variables early in their algebra studies, we may ease students' transition to abstract algebra for those who choose to study it.

The generating-function approach reverses the way we usually think about expanding $(c + h)^x$ and ushers in another initially uncomfortable shift in our thinking. Usually, we would draw Pascal's triangle or invoke the binomial theorem to find the coefficients and the terms of the expansion of $(c + h)^x$. In the preceding probability problem, we accept the expansion as a *given*, courtesy of the CAS, and use the expansion to find the coefficients. Alternatively, we can even use a series of CAS-generated examples of binomial expansions to apply Pascal's triangle by letting the exponents go from 0 to 1 to 2, and so on. Working by hand might lend some insight into how the pattern arises and some motivation for determining the pattern to avoid the tedium of working by hand.

Problem 2: "Making Change"

We next turn from the subject of sports to the subject of money by presenting the problem below from Kahan and Wyberg (in press) and insights contributed by our student David Ehren.

> Can you make exactly five dollars with exactly 100 U.S. coins and no nickels? If so, how?

This problem, which we first heard in a class by Doug McGlathery in the Cambridge Rindge and Latin High School, can be approached with generating functions. In solving the problem above, we model solving it with pennies, dimes, and quarters using trial and error—like most students do—instead of generating functions. Each time we choose one of the 100 coins, we add 1, 10, or 25 to the value, in cents. We choose a coin 100 times, so by expanding $(x^1 + x^{10} + x^{25})^{100}$, we should represent all the possible values. In the expansion, the exponents represent the values and the coefficients represent the number of ways each amount can be made. In the Making Change problem, x is used even more abstractly than in the How Many Games Should You Play? problem. In the Making Change problem, x stands for nothing at all. By calling $(x^1 + x^{10} + x^{25})^n$ a function, we implicitly assume we are evaluating $(x^1 + x^{10} + x^{25})^n$ where n, the number of coins, is taken to be 100.

Before we explore the case of 100 coins, consider the more accessible example of modeling the choice of 2 coins by $(x^1 + x^{10} + x^{25})^2$. The first coin will be called coin A, and the second coin will be called coin B. If we think of $(x^1 + x^{10} + x^{25})^2$ as $(x^1 + x^{10} + x^{25})(x^1 + x^{10} + x^{25})$, the first factor can represent the possible choices of coin A. A penny is represented by x^1; a dime, by x^{10}; and a quarter, by x^{25}. If we multiply x^{10} from the first factor by x^1 in the second factor, we get x^{11}. This product represents choosing coin A as a dime and coin B as a penny. The 11 in x^{11} represents the sum of the values of coin A and coin B. Using CAS to expand $(x^1 + x^{10} + x^{25})^2$, we get $x^{50} + 2x^{35} + 2x^{26} + x^{20} + 2x^{11} + x^2$. The $1x^{50}$ term provides us with two sources of information. The exponent tells us that the sum of the two coins is 50 cents, and the coefficient of 1 tells us that we can get 50 cents in one way, namely, coin A and coin B must both be quarters. The $2x^{11}$ tells us that we can get 11 cents in two ways. One possible way to get 11 cents is if coin A is a dime and coin B is a penny. The other possibility is that coin A is a penny and coin B is a dime. Note that this generating function treats each factor as a separate and distinct coin and that one might therefore think that the coefficients "overcount" the number of possible ways to make certain sums. The expression $x^{50} + 2x^{35} + 2x^{26} + x^{20} + 2x^{11} + x^2$ gives us more information than the number of ways to make certain sums using two coins; it also tells us which sums are impossible. No term appears with a variable part of x^3. This statement makes sense because no combination of two coins that result in a sum of 3 cents exists. Note too that a value for x is not meaningful in this problem. To gain further insight, teachers may want students to expand $(x^1 + x^{10} + x^{25})^3$ by hand and to explain what problem this expression represents and how it does so.

Returning to the case of using $(x^1 + x^{10} + x^{25})^{100}$ to explore the choice of 100 coins, we have exceeded the limits of what we can calculate by hand, so without an advanced CAS this approach is a nonstarter. CAS expands $(x^1 + x^{10} + x^{25})^{100}$ to an expression that begins $x^{100} + 100x^{109} + 4950x^{118} + 100x^{124}$. These terms make some sense. We have one possible way to obtain a value of 100 cents, namely, by making all the coins pennies. If we trade a penny for 1 dime, we have 99 pennies and 1 dime, for a value of 109. The sense in which 100 ways are possible to do so, as the coefficient claims, is that we may trade any of the 100 pennies. We can make 118 cents with 98 pennies and 2 dimes, and we have 4950 ways to choose 2 pennies to replace from the 100. The next term tells us that 100 ways are possible (in what sense?) to make 124 cents (how?).

As we look across the exponents, we notice that they increase by multiples of 3, a pattern that in hindsight makes sense. Any pair of penny, dime, and quarter differs by a multiple of 3 (e.g., a penny and a quarter differ by 24 cents), so any exchange of one coin for another changes the total by a multiple of 3. Moreover, any exponent that appears is 1 more than a multiple of 3. However, 500 is not 1 more than a multiple of 3, so it should not appear. As we proceed we find

$$1290507887523833545964175199713600 x^{499}$$
$$+1744742299936097432911180037836500 x^{502}$$

and the assertion is confirmed. Therefore, the generating function shows that no solution is possible when we use only pennies, dimes and quarters.

The generating-function approach to this problem is structurally similar to the more conventional algebraic approach. Looking for a solution with pennies, dimes, and quarters, we might begin by letting p, d, and q represent the numbers of pennies, dimes, and quarters, respectively. Then

(1) $$p + d + q = 100.$$

and

(2) $$p + 10d + 25q = 500.$$

Subtracting equation (1) from equation (2), we obtain

(3) $$9d + 24q = 400.$$

If d and q are whole numbers, then the left-hand side of step (3) above should be a multiple of 3; but the right-hand side is not, so again no whole-number solutions are possible. The underlying reason again relates to divisibility by 3. If instead of expanding $(x^1 + x^{10} + x^{25})^{100}$ we were to factor out x first, we would obtain $[x(1 + x^9 + x^{24})]^{100} = x^{100}(1 + x^9 + x^{24})^{100}$. The problem then reduces to looking for an x^{400} term in the expansion of $(1 + x^9 + x^{24})^{100}$. Expanding this result with CAS, we observe what we perhaps in hindsight could have predicted—that each exponent is a multiple of 3, so 400 cannot appear. Thus the parallel between the algebraic method and the generating-function one is quite strong.

What if we allow half-dollar and dollar coins? Then we need to look for an x^{500} term in $(x^1 + x^{10} + x^{25} + x^{50} + x^{100})^{100}$ or, using the method of the previous paragraph, an x^{400} term in $(1 + x^9 + x^9 + x^{24} + x^{49} + x^{99})^{100}$. Using CAS to expand either of these expressions, we find that the desired term is present, so with pennies, dimes, quarters, half-dollars, and dollars, we can make five dollars with 100 coins.

We do not want to be like the mathematician who, on being awakened by a fire in his room, sees a bucket near the bathroom sink and says, "Ah yes, a solution exists," and goes back to sleep. If we persist with the generating function, can we find a solution?

Let us make some observations. The first is that we will need half-dollars. In the expansion of $(1 + x^9 + x^{24} + x^{99})^{100}$ all exponents will be multiples of 3, so 400 will not appear as an exponent. Could we do without dollar

Technology Matters: An Invitation to Generating Functions with CAS **159**

coins? CAS demonstrates that an x^{400} term appears in $(1 + x^9 + x^{24} + x^{49})^{100}$, so yes, a solution is possible with pennies, dimes, quarters, and half-dollars. Since, like the 400 we seek, 49 is also 1 more than a multiple of 3, trying combinations with just one half-dollar seems like a good idea. Using 4 half-dollars or 7 half-dollars would also be good choices, but 2, 3, 5, or 6 would not. Why not?

Adding in one half-dollar corresponds to multiplying in one of the x^{49} terms, so we decrease our desired exponent of 400 by 49 and our number of factors by 1 and look for what we can do with the pennies, dimes, and quarters. In the generating-function world, we next seek an x^{351} term in $(1 + x^9 + x^{24})^{99}$. We find one, because $9 \cdot 39 = 351$. If we multiply in the x^9 term 39 times, the 1 term the remaining 60 times, and the x^{24} term no times, we obtain the desired exponent. Furthermore, in the real world, we have also just found a solution, namely, 1 half-dollar, 0 quarters, 39 dimes, and 60 pennies.

By exploring the generating functions, we can find more solutions in this manner, especially if we have some knowledge of modulo, or clock, arithmetic. In fact, by combining patience with number sense, we could find all the solutions and make the case that no others are possible. The divisibility arguments we would make strongly parallel the ones we would make if we approached this problem as a system of Diophantine equations—equations requiring solutions with integral values. Carrying this plan out would be somewhat tedious and reaches beyond the scope of this chapter.

However, the problem can be solved less tediously with a CAS and a small amount of programming. The difficulty we encountered was that the x^{500} term did not let us see which factors had built it up. If instead we expand

$$(p + d^{10} + q^{25} + h^{50} + w^{100})^{100}$$

we should be able to see the exact contributions of each coin (p holds the place for pennies; d, for dimes; q, for quarters; h, for half-dollars; and w, for whole dollars) to the terms with degree 500, if we live that long. But a small programming change, throwing out any terms of degree over 500 before multiplying again, gives a polynomial within a minute (further improvements are possible, such as setting any coefficients to 1 or making the cutoff 500 minus the number of multiplications left), and a term of that polynomial is $14540659943409600 d^{60} h^{50} p^{90} w^{300}$. The coefficient tells us that if we choose the coins in order, we can choose them in many ways. The exponents tell us which coins and how many, namely, 6 dimes (60 cents' worth), 1 half-dollar, 90 pennies, and 3 whole dollars. All together, nineteen terms have degree 500, and so the problem has nineteen solutions.

We can also obtain the 19 solutions described above with a computer program written to exhaust all the possibilities rather than one that uses generating functions. The computer solution yields an exhaustive list of the solutions, but we need to explore the list to gain insight into the problem. Kahan and Wyberg (in press) provide other methods for approaching the coins problem. For example, as we noted above, it is not coincidence that all solutions have 1, 4, or 7 half-dollars. The generating-function is one approach worth using and understanding, but it is not the only one. Some educators would say it is not the best approach, and if forced to choose, we would agree, but we tend to think that looking for one single best approach does not do justice to the mathematics or its teaching.

Summary and Implications for Teaching

In this chapter, we have explored two mathematics problems in which a generating-function approach, aided by CAS, can be useful. We have focused on probability and combinatorics because these applications are among the most accessible of generating functions. Tucker (1984) offers a more extensive treatment of generating functions in combinatorics, and Wilf (1994) presents more information on generating functions in many branches of mathematics. DeMoivre used generating functions to obtain a closed-form expression for the Fibonacci numbers (Tucker, 291–292.) This chapter is an invitation to learn more about generating functions.

This chapter also illustrates how CAS supports work with generating functions. Now that CAS can expand the polynomials for us, if we can represent a problem with a generating function, CAS can help us solve the problem. As the number of games in the tournament-series problem rises, hand expansion becomes less possible. Therefore, before CAS, the generating function quickly became a less efficient choice. For the Making Change problem, a generating function approach is reasonable—only if a CAS is available. The technology changes the efficiency or even the possibility of using generating functions to approach a problem.

Both problems introduced herein were solved using generating functions with help from a CAS, but the goal of this chapter is to focus on ways we can better teach mathematics. We believe that students can use generating functions to make more sense of difficult problems and that this process can foster additional insight into the mathematics they are learning. The coordination of generating-function approaches to these problems, along with other approaches, reveals underlying mathematical structure and prepares students for abstract algebra. Students may share Tucker's (1984, p. 232) pleasure in seeing how "the algebraic techniques automatically do the combinatorial reasoning for us" with generating functions. Using generating functions in these problems illustrates how the concepts of variable and function and the understanding of probability and combinatorics are extended and enhanced.

REFERENCES

Coxford, Arthur F., James T. Fey, Christian R. Hirsch, Harold L. Schoen, Gail Burrill, Eric W. Hart, Ann E. Watkins, Mary Jo Messenger, and Beth Ritsema. *Contemporary Mathematics in Context: A Unified Approach, Course 1, Part B.* Chicago: Everyday Learning, 1997.

Dienes, Zoltan P. *Building Up Mathematics*. London: Hutchinson Educational, 1960.

Finzer, William. Fathom Dynamic Statistics Software. Version 1.1. Berkeley, Calif.: Key Curriculum Press, 2001. Software.

Kahan, Jeremy A., and Terrence R. Wyberg. "Mathematics as Sense Making." In *Teaching Mathematics through Problem Solving: Grades 6–12*, edited by Randall Charles and Harold Schoen. Reston, Va.: NCTM, in press.

Kline, Morris. *Mathematical Thought from Ancient to Modern Times*. New York: Oxford University Press, 1972.

National Council of Teachers of Mathematics (NCTM). *Principles and Standards for School Mathematics*. Reston, Va.: NCTM, 2000.

Tucker, Alan. *Applied Combinatorics*. 2nd ed. New York: John Wiley & Sons: 1984.

Usiskin, Zalman. "Conceptions of School Algebra and Uses of Variable." In *The Ideas of Algebra K–12*, 1988 Yearbook of the National Council of Teachers of Mathematics, edited by Arthur F. Coxford and Albert P. Shulte, pp. 8–19. Reston, Va.: NCTM, 1988.

Wilf, Herbert S. *Generating Functionology*. 2nd ed. San Diego, Calif.: Academic Press, 1994.

Activity 5

Given the quadrilateral with vertices $A(-5, 2)$, $B(11.3, 7.1)$, $C(16.4, 5.0)$ and $D(0.1, -0.1)$.

    (a) Show that ABCD is a parallelogram.

    (b) Are the diagonals perpendicular? Show how you know.

    (c) Show that the diagonals bisect each other.

Solution (a)

- Define slope (a1, b1, a2, b2) $= \dfrac{b2 - b1}{a2 - a1}$ Done
- slope(-5,2,11.3,7.1)   51/163
- slope(11.3,7.1,16.4,5)   ⁻7/17
- slope(16.4,5,.1,⁻.1)   51/163
- slope(.1,⁻.1,⁻5,2)   ⁻7/17

Since both pairs of opposite sides have the same slope, the quadrilateral is a parallelogram.

Solution (b)

- slope(⁻5,2,16.4,5)→md1   15/107
- slope(11.3,7.1,.1,⁻.1)→md2   9/14
- md*md2 = ⁻1   false

Since the product of the slopes of the diagonals is not –1, the diagonals are not perpendicular.

Solution (c)

- $\left\{\dfrac{a1+a2}{2} \ \dfrac{b1+b2}{2}\right\}$ →mdpt(a1,b1,a2,b2)   Done

- midpt(⁻5,2,16.4,5)   (57/10 7/2)
- midpt(11.3,7.1,.1,⁻.1)   (57/10 7/2)

Since the diagonals have the same midpoint, they bisect each other.

CHAPTER 10

To CAS or Not to CAS?

James E. Schultz

WITH COMPUTER algebra systems (CAS) capable of performing virtually all of the exact computations in courses through calculus and readily available in relatively inexpensive handheld versions, considerable controversy has arisen over their use. In response to the question "To CAS or not to CAS?" I reflect on the implementation of earlier technologies and give examples of appropriate and inappropriate uses of CAS. I also describe a particular example in detail and include the outcomes as they relate to students. Finally, I reexamine my opening question, "To CAS or not to CAS?" in light of the examples explored in this chapter.

Since handheld calculators first appeared in the 1970s, people outside the school classroom have used them to perform the four basic arithmetic operations. For the most part, calculator skills have replaced paper-and-pencil skills, but mental arithmetic remains important. Problem solving, estimation skills, and the basic operations are a natural part of computation with calculators and encourage students to ask and answer two questions: (1) What operation(s) should be performed? and (2) Is the answer reasonable? Similarly, CAS may be the technology that supplants numerical, algebraic, calculus, and other computations in the future, but such an occurrence appears to be slow in coming.

A comparison of the NCTM Standards of 1989 and 2000 suggests that the 1989 *Curriculum and Evaluation Standards for School Mathematics* (*Curriculum and Evaluation Standards*) offers strong support for graphing calculators, the most widely available tool for school mathematics at the time of writing. The 2000 *Principles and Standards for School Mathematics* (*Principles and Standards*) only mildly supports CAS, currently the most powerful tool available (Waits 1999). Significant CAS activity occurs at the middle and high school levels, but this activity exists primarily in countries other than the United States, particularly in Australia and Austria. See Ball (2001) and Herget and colleagues (2000).

The term "CAS" in this chapter refers to the standard—often handheld—versions of CAS rather than to CAS specifically enhanced for pedagogical purposes. Although these machines often have graphing utilities available, this chapter focuses on symbolic manipulation, specifically addressing those manipulations that cannot be done on a graphing calculator.

Perspectives on Handheld Technology

One of the first uses of handheld technology to gain acceptance centered on tasks formerly done using tables, such as finding square roots, logarithms, and trigonometric functions. In years past, students learned to use tables—in combination with other techniques—to compute. For example, $\sqrt{80}$ could be obtained by multiplying $4\sqrt{5}$, where $\sqrt{5}$ was obtained from a table or was memorized. Similarly, log 124 could be obtained as 2.0934 by writing 124 as 1.24×10^2, giving a "mantissa" of .0934 and a "characteristic" of 2. Also, sine 23° 30′ could be obtained by interpolating between sine 23° and sine 24°. The three techniques—however mathematically interesting or uninteresting they may have been—are now essentially subsumed by the calculator.

In light of this evolution, what *should* well-prepared students know about $\sqrt{80}$ in this technological age? Four things come to mind: (1) students should know the concept of square root and when to use it in a problem-solving setting; (2) students should be capable of using the calculator to find $\sqrt{80}$ correctly; (3) students should reason that $\sqrt{80}$ is slightly less than $\sqrt{81}$, and be able to estimate that the number produced by the calculator is reasonable; and (4) students must know the result of $\sqrt{81}$ *without* a calculator. The scenario might unfold as follows:

> "If a square room has an area of 80 square feet, how long is each side?" Representing this problem uses the square root concept, so we need $\sqrt{80}$. The calculator gives about 8.944, which seems reasonable, since we expect an answer slightly less than $\sqrt{81} = 9$.

Analogous conditions are illustrated here for solving $60(1.02)^x = 70$ using CAS. A scenario for the problem above might develop as follows:

> "If the population of a country is about 60 million and growing at a rate of 2% per year, in how many years will it reach

70 million?" This situation can be represented by an equation involving an exponential function with a multiplier of 1.02, since each year the population will be 102 percent of that of the previous year. Write the equation $60(1.02)^x = 70$ to model the situation. Using CAS to solve $60(1.02)^x = 70$ for x gives 7.78 (to two decimal places). To check, observe that the growth in the first year will be 2 percent of 60, or 1.2. Thus an overall growth of 10 million should occur in about 8 years, so the answer of 7.78 years seems reasonable.

Appropriate Use of CAS

The examples below illustrate a variety of ways that CAS can help students focus on solving problems and learning concepts.

Numerical

1. Evaluate $\dfrac{52!}{13!\ 39!}$ in a probability problem about playing cards.

2. Factor 11,111 into primes when exploring repeating decimals (to see which fractions will repeat in blocks of five digits).

Algebraic

1. Compute $a^6 \cdot a^2$, $\dfrac{a^6}{a^2}$, and $\dfrac{a^2}{a^6}$ to learn about laws of exponents.

2. Expand $x + y$, $(x + y)^2$, $(x + y)^3$, $(x + y)^4$, $(x + y)^5$, …to see a relationship with Pascal's triangle.

3. Find the results of subtracting 5 from each side of the equation $5x = 10$ (a common error) when learning how to solve equations (Kutzler 1999).

Calculus

1. Find the derivative of $l(x) = \dfrac{4}{\sin x} + \dfrac{2}{\cos x}$ in a max/min problem involving the length of a rod (Edwards and Penny 1998).

2. Evaluate $\lim \Sigma \dfrac{1}{n} \cdot \dfrac{k^2}{n^2}$, for $k = 1$ to n as $n \to \infty$, to show the definition of $\int_0^1 x^2\, dx$. CAS gives 1/3 as an exact value of this limit of upper sums.

This idea comes from Michael Waters, who developed it in a high school classroom in response to a question from a student.

CAS, Not Always the Most Appropriate Approach

The examples below illustrate a variety of instances in which use of CAS is not necessarily the most appropriate approach.

Numerical

1. Evaluate $\dfrac{52!}{51!\,1!}$ in a probability problem about playing cards. By definition of factorial, this result can be seen to be 52.

2. Factor 10,000 into primes when exploring repeating decimals.

Algebraic

3. Compute $a^6 \cdot a^2$, $\dfrac{a^6}{a^2}$, and $\dfrac{a^2}{a^6}$ *after* laws of exponents have been learned.

4. Expand $(x + y)^2$. Students should learn this basic fact as a referent.

5. Solve the equation $5x = 10$ after methods of solving linear equations are learned.

Calculus

6. Find the derivative of $y = \cos x$ when x is 0. The answer is apparent from the graph.

7. Evaluate $\displaystyle\int_{-2}^{2} 8x^3\,dx$. This is an odd function, so by symmetry the integral is 0.

A comparison of the problems in the "appropriate" versus "*not* the most appropriate" categories shows that some computations may be appropriate at a certain time or with a specific group of students, but they may not be appropriate under different conditions. The next example describes in detail an example of effective CAS use with students enrolled in a class that has, as a prerequisite, only minimal knowledge of algebra.

An Extended Example of Appropriate Use of CAS

One application of mathematics that is important for students to learn is the payment formula for installment loans. Because of the popularity of credit cards, car loans, and home mortgages, most students eventually encounter this application, which is a favorite of mine because I saved over $3000 by uncovering a bank error on a home mortgage. Although such errors are uncommon, understanding the mathematics underlying loans is important for almost everyone. Consider this example:

> If the interest rate on a home mortgage loan is 9% per year, how much is saved by repaying a loan of $100,000 in 15 years instead of 30 years?

With the help of CAS, students can learn that the answer is more than $100,000, which is more than the original loan amount!

Schultz and Noguera (2000) describe how the problem above was used successfully as a culminating activity with students enrolled in a low-level college mathematics course. Here I outline ways to use this example with secondary school students. A number of activities—explained below—can lead to students' solving problems using the payment formula.

1. Students learn to write equations to model linear and exponential situations.
2. Students approximate solutions to such equations by trial and error using first calculators and then spreadsheets.
3. Students learn how to solve simple linear equations without technology.
4. Students learn how to solve linear equations using CAS.
5. Students learn how to solve exponential equations, giving approximate solutions, using CAS. The role of logarithms in the solution might be only casually mentioned.
6. Students learn to solve problems with the payment formula using CAS.

Items 1–3 above represent "white box" methods—for students who have some understanding of what the technology is doing to obtain the result. Similarly, items 4–6 above represent "black box" methods—for students who have no understanding of what the technology is doing to obtain the result. See Buchberger (1989) for a discussion of the "white box, black box" principle.

The focus of solving the mortgage problem above is similar to the focus of the population problem involving exponential functions and should include the following steps:

- Modeling the situation by writing an equation
- Using the technology—in this case CAS—to solve the equation
- Using estimation and previous results to check the answer for reasonableness

Each of these steps is developed in greater detail below:

Modeling the situation. Modeling the situation can be given higher priority so that the focus is on setting up equations to model situations instead of solving given equations. Students set up linear equations to model simple interest and exponential equations to model compound interest. To lead up to the payment formula

$$\text{payment}(a, r, n) = \frac{a \cdot r \cdot (r+1)^n}{(r+1)^n - 1},$$

where *a* is the amount of the loan, *r* is the interest rate for the payment period, and *n* is the number of payments, students begin by making spreadsheets. They use guess-and-check methods to determine the payments required to pay off a balance, such as a $500 credit card balance with 1 percent monthly interest in four months, as shown in figures 10.1, 10.2, and 10.3.

Next, students use CAS to show that the result of beginning with $500, adding 1 percent monthly interest, and subtracting a payment of *p* dollars leads to a balance of 520.30 – 4.06*p* after four months. They enter the original amount of 500. They can use the constant feature of the machine to

B3		f_x =B2 + C2 - D2	
A	B	C	D
1 Months	Balance	Interest	Payment
2 0	$500.00	$5.00	$127.00
3 1	$378.00	$3.78	$127.00
4 2	$254.78	$2.55	$127.00
5 3	$130.33	$1.30	$127.00
6 4	$4.63		

Fig. 10.1. A payment of $127 is too small to pay off a $500 balance in four months.

B3		f_x =B2 + C2 - D2	
A	B	C	D
1 Months	Balance	Interest	Payment
2 0	$500.00	$5.00	$130.00
3 1	$375.00	$3.75	$130.00
4 2	$248.75	$2.49	$130.00
5 3	$121.24	$1.21	$130.00
6 4	-$7.55		

Fig. 10.2. A payment of $130 is too large to exactly pay off a $500 balance in four months.

	A	B	C	D
		B3	f_x =B2 + C2 - D2	
1	Months	Balance	Interest	Payment
2	0	$500.00	$5.00	$128.14
3	1	$376.86	$3.77	$128.14
4	2	$252.49	$2.52	$128.14
5	3	$126.87	$1.27	$128.14
6	4	$0.00		

Fig. 10.3. A payment of $128.14 pays off a $500 balance in four months

repeatedly multiply the previous answer by 1.01 and subtract the monthly payment of p. The respective balances are listed below:

$$500$$
$$505 - p$$
$$510.1 - 2.01p$$
$$515.2 - 3.03p$$
$$520.3 - 4.06p$$

Notice that with a monthly interest rate of 1 percent, the multiplier is 1.01, reflecting that the new amount, including interest, is 101 percent of the previous amount, as in the example discussed previously. Setting the final balance equal to 0, then solving for p gives a payment of $128.14.

After working several examples, students should be able to generalize their findings, thus deriving the payment formula. A more rigorous derivation of the payment formula, using the sum of a geometric series, can also be presented to students if appropriate.

Using CAS to solve the equation. Using CAS to solve the equation once the formula is determined involves defining it in the students' machines, where it can be called on when needed. For example, using the payment formula given previously for an original amount of $100,000 at a yearly interest rate of 9 percent (0.75% monthly) for thirty years (360 months) yields a monthly payment of $804.62.

With 360 payments of $804.62 resulting in a total payback of $289,663.20, the total interest paid is $189,663.20. Similarly, solving with a CAS finds the monthly payment for fifteen years to be $1014.27 with a total payback of $182,568.60, so the total interest is $82,568.60. Therefore, paying an additional $210 per month saves $107,094 in interest! Thus, increasing the payments by only about 25 percent saves well over 50 percent of the interest. This valuable lesson and others like it are far more accessible when

CAS are available. Moreover, a handheld CAS can empower consumers when they shop for loans and compare finance charges.

Using estimation and previous results to check the answer for reasonableness. A recommendation that appeared in the NCTM's 1989 *Curriculum and Evaluation Standards* (p. 98) can be extended to using estimation and previous results to check the answer for reasonableness when solving problems with the payment formula. The ideas expressed in that *Standards* document reflect the belief that students should be provided with ways to check the reasonableness of computations, including methods for avoiding errors in placing the decimal point in such "basic fact" problems as $0.1 + 0.1 = 0.2$ and $0.1 \times 0.1 = 0.01$. By learning a few examples as *referents* and exploring how changing the examples affects the results, students can apply this principle to solving problems with the payment formula. For example, students who remember how to solve the previously discussed home mortgage problem can use it as a *referent* for solving other similar problems, such as those below.

- If amount of the loan was $50,000, the payment would decrease proportionately.
- If the yearly interest rate was 10 percent, the payment should increase, but *not* proportionately, because exponents are involved.

Impact on Students

Schultz and Noguera report (2000) that using CAS in a low-level course allowed students to investigate a wide range of problems involving linear, exponential, quadratic, rational, radical, and combinatorial functions, as well as problems similar to the extended example above. The following outcomes were realized as a result of this study:

- Greater focus on problems relevant to the students' lives
- Greater emphasis on technology skills, less on paper-and-pencil skills
- Greater emphasis on students' thinking, less on routine computations
- Greater attention to modeling situations instead of solving given equations
- Greater attention to number sense and estimation

The benefits listed above were enhanced by students' previous use of technology—already a part of their lives. The gradual immersion into technology, culminating with CAS, enabled the students to solve interesting and significant problems. In sharp contrast with feedback from this audience in traditional courses, 95 percent of the students enjoyed the course and perceived that it was valuable (Schultz and Noguera 2000).

Conclusion

In this chapter, I have illustrated that CAS can play a positive role in the teaching of mathematics, even at lower levels, although the approach is not always the most appropriate. From a current perspective, my original question, "To CAS or Not to CAS?" is *not* the question. The real questions today are *"When* to CAS?" and *"With whom* to CAS?"

REFERENCES

Ball, Lynda. "Computer Algebra Systems in Schools Curriculum, Assessment, and Teaching Project." Melbourne: University of Melbourne, 2001. www.edfac.unimelb.edu.au/DSME/CAS-CAT/.

Buchberger, B. *Why Should Students Learn Integration Rules?* RISC-Linz Technical Report no. 89-7.0, University of Linz, Austria, 1989.

Edwards, C. H., and David Penny. *Calculus and Analytic Geometry*. 5th ed. Upper Saddle River, N.J., 1998.

Herget, Wilfried, Helmut Heugl, Bernhard Kutzler, and Eberhard Lehmann. "Indispensable Manual Calculation Skills in a CAS Environment." 2000. www.kutzler.com (6 Sept. 2001).

Kutzler, Bernhard. "The Algebraic Calculator as a Pedagogical Tool for Teaching Mathematics." 1999. www.kutzler.com (6 Sept. 2001).

National Council of Teachers of Mathematics (NCTM). *Curriculum and Evaluation Standards for School Mathematics*. Reston, Va.: NCTM, 1989.

———. *Principles and Standards for School Mathematics*. Reston, Va.: National Council of Teachers of Mathematics, 2000.

Schultz, James E., and Norma Noguera. "High Level Technology in a Low Level Mathematics Course." *The International Journal of Computer Algebra in Mathematics Education* 7(1)(March 2000): 25–32.

Waits, Bert K. Circulated e-mail message, March 21,1999. Used with permission.

CHAPTER 11

Task Design in a CAS Environment: Introducing (In)equations

Nurit Zehavi
Giora Mann

THE IDEA of using computers to perform symbolic rather than numerical calculations led to the development of computer algebra systems (CAS) in the early sixties (Harper, Wooff, and Hodgkinson 1991). The availability of CAS for microcomputers in the mid-eighties attracted mathematics educators to the possibility of using this technology in the classroom. Small, Hosack, and Lane (1986) were among the first to use tasks that encourage students to apply techniques that they already understand in simple cases to complex problems wherein a symbolic tool can cope with difficult manipulations. Clearly, such tasks broaden the scope of the traditional curricular goals. At the same time, mathematics educators started to explore new teaching methods that use CAS to enhance teaching and learning mathematical topics, allowing students to concentrate on conceptual aspects as they learn the topics. Early CAS-based studies by Heid (1988), Judson (1990), and Palmiter (1991) indicated that resequencing of the content so that the concepts are taught before the manipulation skills was effective for achieving a greater understanding of concepts without decreasing mastery of manipulation skills.

As more studies were conducted, more questions were asked. What tasks involving CAS engage students in conceptual aspects? How do we assess the quality of these tasks? What is the role of CAS in emerging curricula? In response to these questions, Pozzi (1994) found that students who do not fully comprehend the output produced by CAS develop informal— and possibly erroneous—ideas about what the computer does. Therefore, using CAS might necessitate a better conceptual understanding of the algebraic manipulations. In an attempt to describe the unity and diversity of mathematical thinking in general terms, Aspetsberger, Fuchs, and Schweiger (1997) explored the notion of *fundamental ideas* in a CAS environment. They suggested a framework of curricular "islands" in which fundamental ideas meet topics, assuming that sharing the same ideas in different topics can help students extend the meaning of mathematical concepts. Their framework can be viewed within a global theory of *webs of meaning* developed by Noss and Hoyles (1996). The idea of *webbing* is meant to convey the presence of a structure that learners can draw on and *reconstruct* for support, in ways that they deem appropriate, to increase their understanding of a mathematical topic. With CAS, which is in fact a mathematical *instrument*, learning to use the instrument to construct mathematical meaning is both complex and insightful for researchers and curriculum designers (Guin and Trouche 1999). As more mathematics teachers became familiar with CAS, a new area surfaced, namely the design of CAS-written support materials.

The MathComp Project

The MathComp (mathematics on computers) project began in 1995 in the Science Teaching Department at the Weizmann Institute of Science with the aim of integrating CAS into teaching to improve the learning of mathematics. In the first stage of the project we considered the future needs of teachers' preparation and their involvement in the process. Thus we designed and implemented professional development courses for teachers, familiarizing the teachers with the Derive software by exploring problems at *their own level* (MathComp team 1998). Only after they had experienced doing mathematics with CAS did we challenge them to create problems for students (Zehavi 1996a, 1997a). During the course, the teachers became more independent and creative while incorporating CAS into their professional lives. Moreover, we encouraged them to rethink curricular and didactical aspects of mathematics learning and teaching, and we learned with them.

In 1996 we began preparing CAS-based mathematical workbooks for junior high school (MathComp team 1999), and in 1999 we began developing material for the upper high school grades. The learning units in the workbooks complement the current syllabus. They are designed in the format of interactive workshops in the computer lab so that students learn through investigation and discussion. The basic assumption in planning the MathComp units is that it is presently too early to implement a full CAS curriculum; thus students learn algebra with paper and pencil in class and use CAS in the lab to broaden learning opportunities and to promote greater mathematical understanding. Each of the units includes a variety of *tasks* aimed at improving skills, understanding concepts, and investigating problems. These tasks create networks of mathematical connections, add depth to the subject at hand, and expose students to new mathematical experiences.

The process of producing commercial versions of the MathComp materials took three years. The first year was devoted to the design and production of pilot versions of the material. The pilot trials took place in classes of MathComp team members or in classes in which the teachers volunteered to use the material and invited team members to be present. On the basis of the findings from the pilot work, we prepared a limited edition of experimental materials. We invited teachers to participate in professional development workshops where the materials were presented. Some teachers decided to implement the learning units in their classes. Among them we sought volunteers to document and report their experiences. In the workshops that followed we discussed the findings and asked teachers for their comments, suggestions, and requests. Consequently, we revised the material for the commercial version. Classroom experience and teachers' workshops are part of the ongoing formative development of the materials.

During this long-term development-and-implementation process, we learned to formalize our discussions in terms of the following *principles:* learning environment, teaching techniques, and role of the teacher.

Learning environment. The mathematical assistant, as Derive is called, serves as a learning environment in which students' work, CAS performance, and students' reflections are intertwined.

Teaching techniques. The teaching techniques involve consideration of both up-to-date technology and modern approaches to student learning, for example, (*a*) the balance between the knowledge conveyed by the teacher and its construction by the student, and (*b*) the emergence of sociomathematical norms.

Role of the teacher. The role of the teacher who teaches with modern technology is very complex; thus the structure and options of a computer-based learning unit should initially be made clear to the teacher at a global level.

We set out to develop tasks that, in our view, could not be dealt with practically and effectively without CAS. The tasks were intended to highlight the fundamental ideas within a mathematical topic. Moreover, the tasks extended the topic at hand by connecting it with previously learned topics or with topics that would be learned at a later stage. In this way, the tasks served to develop students' deeper understanding of the topic at hand.

Since our goal is to improve learning, we started by identifying persistent difficulties in current practice and then selected those difficulties for which we expected the technology to be helpful. By the term *difficulties*, we mean *known misconceptions*, such as students' "seeing an equation as two equal expressions." In each topic, we began to develop new tasks designed to overcome the difficulties specific to it. The formative development process not only helped shape the tasks but also produced new tasks that extended the topic at hand, thereby providing continuity. Students and teachers gave us feedback on the basis of their reactions to the various iterated versions of the tasks. In this chapter, we describe two units from the grade-8 workbook that deal with the notion of (in)equations. We present the rationale and the consequent task design. We describe in detail some parts of the formative development that took place in the classes and in teachers' workshops and discuss the curricular implications.

Rationale

In the local mathematics curriculum for grade 8, equation and (in)equation solving play a central role. Designing CAS-based tasks for teaching the symbolic manipulations that CAS are designed to replace poses an intriguing dilemma for educators. Influenced by the *principle of selective construction* (Tall 1994), we allow students to use the automatic power of the software when equation-solving procedures are *not* the object of our teaching. However, when we want students to learn the procedures, we use the software differently. To enable students to focus on a chain of equivalence transformations that change a given (in)equation into a solution statement, we designed diagnostic learning activities using Derive (Zehavi 1996b, 1997b). The basic pedagogical strategy implemented in these activities is replicating and completing examples. By using partially solved examples and reconstructing (in)equations from solutions with CAS, students maintain global evaluation of the algebraic transformations.

Before teaching symbolic manipulations of equations, we must first address a basic conceptual difficulty. Understanding the conceptual structure of (in)equations requires a flexible view of symbolism as a process and a concept. Students at the beginning of their algebra learning tend to perceive an equation of the form $3x - 4 = 2x - 5$ as a statement about two equal arithmetic processes; in fact, the *results* of the two processes are the same. Thus students have genuine difficulty in interpreting $3x - 4 = 2x - 5$ as the question "What is the set of all numbers that will produce a true statement when substituted in the given expressions?"

The introduction of the concept of *equation* has always been problematic because of the complexity of the concept and because of the various uses of the equals sign (=) in classical mathematics. For example, young students encounter "=" for the first time in problems like 3 + 2 = __. In the preceding problem, students first construct the *execute* meaning of the equals sign, that is, "Calculate the sum of 3 and 2 and write it on the right-hand side of the equals sign." A second meaning appears when students are asked whether 3 + 2 = 5 is a true statement and they must construct the *logical* meaning. A third meaning is the *definitive* meaning, for example, let $x = 5$ when substituting in an algebraic expression or open phrase. When students encounter the "=" in algebra as a component of an equation, they only naturally interpret this sign according to its logical or definitive meaning.

However, we want students to use a fourth, integrated meaning of the equals sign in such problems as "Find all the substitutions for which $3x + 2 = 5x - 7$ is a true statement." A CAS may help to construct this fourth meaning of "=" by applying the Solve command to the equation as a whole. Whereas substitution operates on each side of the equation separately, Solve operates on the new object consisting of both sides and the "=" sign. Thus activities involving Solve may help to overcome the misconception of seeing the equation as two equal expressions. For this conceptual understanding to materialize, we planned to introduce Solve at certain stages of the learning tasks in which algebraic representation is central. Our strategy was to design tasks in which the emphasis is on constructing (in)equations and on facilitating conceptual understanding. In the following we present a sequence of tasks from two units in the eighth-grade MathComp book (MathComp team 1998), "Riddles with Expressions" and "Warring Expressions."

Task Design

Task 1: Expressions and numbers

Two expressions, $230x + 82$ and $2194 - 34x$, are given. Draw three integers from –20 to 20 using the command −20 + RANDOM(41), evaluate the two given expressions for each number, and compare the two results. If you do not "hit" a number for which both results are equal, then substitute numbers of your choice until you get equal results. Try to reach the goal in a small number of steps.

The solution to this riddle (see **fig. 11.1**) is 8. Some students may be lucky enough to get 8 as one of their random numbers.

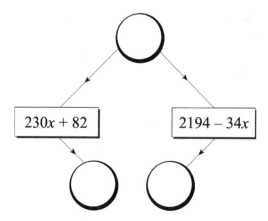

Fig. 11.1. Expressions and numbers

The activity is repeated for other pairs of expressions. As students search for the solution, they should consider the difference between the two results in each subsequent step. This strategy is even more helpful when the solution is outside the given domain for the random numbers or when the solution is not an integer, as in the two expressions

$$2x + 8.2$$

and

$$-\frac{2}{3}x - \frac{1}{3}.$$

A surprise occurs for the pair $6(6-x)+x-16$ and $20-5x$ because any number substituted gives equal results!

The task above is designed for implementation before the students have formally learned about solving equations or after only a brief introduction. The discussion of the class activity is followed by "revealing the secret" that the software can not only evaluate expressions for the numbers substituted but also find the number(s) for which the two results are equal. To find the number(s) using the software, students must write the equals sign between the two expressions and apply the command Solve to the equation. The output is presented on the screen in the following form:

$$2 \cdot x + 8.2 = -\frac{2}{3} \cdot x - \frac{1}{3}$$

$$\text{SOLVE}\left(2 \cdot x + 8.2 = -\frac{2}{3} \cdot x - \frac{1}{3}, x\right)$$

$$x = -3.2$$

Interpreting such output requires understanding the formal algebraic notation. Students are instructed to substitute numbers for x in the equation to obtain a true statement for the solution and a false statement otherwise. This substitution leads the way to using the truth set notation $\{-3.2\}$. Furthermore, applying Solve to the examples that the students solved previously without the software helps them develop trust in the automatic mechanism for solving equations and become acquainted with the specific syntax. For example, where the truth set of the (in)equation is the universal set, the output says "true," but where the truth set of the (in)equation is the empty set, the output says "false."

Next students are given similar tasks and are encouraged to choose the methods they use. The didactical significance of these tasks is that they combine synthesis and analysis processes on equations. The substitution of numbers into two expressions separates the two components of the equation, whereas the solve procedure integrates the two expressions to form a composite entity by operating on this new entity, namely, the equation.

Task 2: Invent a riddle

Next students are asked to complete or invent their own riddles (see **fig. 11.2**). The figure suggests that the results for both sides should be equal.

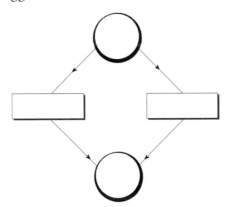

Fig. 11.2. Invent a riddle

Several approaches, ranging from arithmetic to formal algebra, can be applied in constructing a riddle, but we especially welcome the use of the graphical representation. Detailed findings from classroom experience involving this task are described in the following section.

Task 3: Warring Expressions—a game

The unit "Warring Expressions" was designed to deepen students' conceptual understanding of the structure of (in)equations and enhance their expressiveness in algebraic language. The teaching strategy involves students' creating a game situation that motivates them to analyze the game with the various mathematical tools they have already acquired.

> Two expressions, $2(x + 6) - 29$ and $7x + 3(1 - x)$, are given. Draw an integer from -20 to 20, evaluate the expressions for this number, and decide on the winning expression according to the higher numerical result. Draw nine more integers. Which is the winning expression? (In an instance of a tie, neither expression wins.)

Instructions for constructing and playing the game are given. The lines "authored" by the users (see **fig. 11.3**) are aligned to the left, and the results of "simplification" by the software appear next. The third line is a list of random numbers obtained by simplifying the second line. In the following lines the "drawn" numbers are substituted in the given expressions and then organized in a table for finding the winner. The students are asked to repeat the game several times.

```
r := -20 + RANDOM(41)
LIST := [r, r, r, r, r, r, r, r, r, r]
                    LIST := [-2, 12, 1, 17, -20, 19, 7, -11, -17, 6]
VECTOR(2·(x + 6) - 29, x, [-2, 12, 1, 17, -20, 19, 7, -11, -17, 6])
                    [-21, 7, -15, 17, -57, 21, -3, -39, -51, -5]
VECTOR(7·x + 3·(1 - x), x, [-2, 12, 1, 17, -20, 19, 7, -11, -17, 6])
                    [-5, 51, 7, 71, -77, 79, 31, -41, -65, 27]
```

$$\begin{bmatrix} 2\cdot(x+6)-29 & -21 & 7 & -15 & 17 & -57 & 21 & -3 & -39 & -51 & -5 \\ 7\cdot x+3\cdot(1-x) & -5 & 51 & 7 & 71 & -77 & 79 & 31 & -41 & -65 & 27 \end{bmatrix}$$

Fig. 11.3. Constructing and playing the Warring Expressions game

In the game shown, the second expression won 7:3. The accumulated findings of repeated games caused the students to become involved in the unfairness of the game. In the following task they are required to analyze the issue of fairness from the mathematical perspective. This analysis can be done arithmetically, algebraically, and graphically or can be approached from a probability point of view. No specific direction is given. This task revolves around sociomathematical interaction. Consequently, students are challenged to suggest modifications of the game to make it fair. In pilot trials, we observed that students tended to concentrate on numerical arguments when planning a fair game, and as a result, many students became confused. Therefore, we decided to help them achieve the goal by presenting a graphical representation related to the game (see **fig. 11.4**, left-hand side).

The graphical representation is constructed by using Derive's TABLE syntax to get a list of ordered pairs. We hoped it would help students see the two expressions as two parts or sides of one (in)equation, but this strategy was not helpful. Thus we added an arithmetic-algebraic representation in which the (in)equation is explicitly provided, and we used the VECTOR notation of the software to obtain a list of elements to evaluate the inequality for integers from –20 to 20 (see **fig. 11.4**, right-hand side.) The interpretation of the output in mathematical language is that the true-false statements are the result of simplification by the software of numerical statements obtained by substitution of numbers, illustrating the set of truth-values of the inequality in the given domain. **Figure 11.4** combines the main elements of the two representations displayed on the screen. To help students see a unified picture of the game situation, teachers should encourage them to explain the connection between the two representations.

The aim of the unit described above is to integrate topics and fundamental mathematical ideas that students have already learned or will learn. The unit has links with such topics as operations on signed numbers; evaluation of expressions for specific numbers; simplification of expressions; construction of (in)equations from two expressions; truth-set; the graphical representation of expressions; and elementary concepts of probability and statistics, such as frequency.

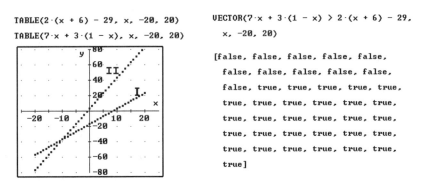

Fig. 11.4. *Graphical and algebraic representations of the unfair game*

Classroom Experience

During the 1997–1998 school year, we implemented the experimental version of the two units, "Riddles with Expressions" and "Warring Expressions," in six eighth-grade classes containing a total of 166 students. All six classes were using Derive for the first time. The teachers documented the events that took place while students were working and collected students' notebooks after each period. The findings regarding the tasks described are discussed below.

Expressions and numbers

The students described previously were engaged in solving riddles (see **fig. 11.5**). Three main approaches to dealing with this task were observed in the six classes.

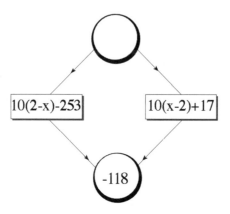

Fig. 11.5. Complete the riddle

An arithmetic/mental algebra approach. The arithmetic/mental algebra approach to solving the riddle above (see **fig. 11.5**) was chosen by approximately 35 percent of the students who participated. Students calculated, in reverse, the solution of one equation, expression = –118, with or without manipulating the expression algebraically and obtained the solution $x = -11.5$. Some students substituted the solution –11.5 in the second expression for verification, and others solved the second expression equalized to –118.

An algebraic/synthesis approach. The algebraic/synthesis approach to solving the riddle above (see **fig. 11.5**) was chosen by approximately 55 percent of the students who participated. Using the computer, students solved algebraically one equation, expression = –118, and substituted –11.5 in the other to verify that both equations obtained the same result, –118. Some students solved two such equations, and a few of them also substituted –11.5 to further convince themselves of the validity of the result.

A formal/analytic approach. The formal/analytic approach to solving the riddle above (see **fig. 11.5**) was chosen by approximately 10 percent of the students who participated. Using the computer, students wrote and solved an equation consisting of the two given expressions. However, most of the students also substituted the solution in one or two expressions.

For the next riddle, only two expressions were given to the students, $4500 + x$ and $1000x + 3501$. We intentionally chose the preceding expressions so that the solution is the number 1, a solution that is easy to obtain by "looking" or guessing. About 25 percent of the students added the equality sign and solved the equation by guessing. Approximately 25 percent of the students wrote and solved the equation using the Solve command. As before, about 50 percent of the students—primarily the same students who used the foregoing algebraic/synthesis approach—first considered each of the expressions separately, substituted numbers, and compared the results. Only after completing the missing numbers in the figure did some students write an equation and solve it to verify their answer. For those students, the Solve command played a different role, that of checking their numerical answer.

Invent a riddle

When students were asked to invent their own riddles, this use of the Solve command became more frequent. The teachers reported that most of their students chose a number and one expression, substituted the number in the expression, wrote another expression, and substituted their chosen number in the second expression. On the basis of the gap between the two results, students "improved" the second expression until they obtained equal results. Many of them used the software to solve equations, thus maintaining control of their work. However, a few students started to invent a riddle by creating an equation and solving it. If the solution did not look "nice," they repeated the procedure.

Two teachers described special inventions in the classroom. In one class, a student chose to first substitute the number 10 in the expression $3(x-2) + 5$. By evaluating the preceding expression for $x = 10$, the result was 29. The student made some changes in the first expression to obtain the second expression, making sure that substituting 10 would yield 29. The student came up with the expression $3(x + 2) - 7$. Next, the student used the computer to solve the equation $3(x - 2) + 5 = 3(x + 2) - 7$, and the computer returned "true"! The teacher was asked for help in explaining what "went wrong." The teacher explained that in fact both expressions simplify to $3x - 1$ and reminded the student that by "true" the software means that all the numbers are in the truth set of the equation. However, the students kept asking, "How does one get such an equation without planning?" In a similar situation in the second class, a student and the teacher posed the preceding question to the whole class. The teacher found it difficult to explain what really occurred. "If the coefficients of the variable x are the same in both linear expressions and the expressions give the same result when evaluated for a specific number, they must simplify to the same expression." The teacher then decided to base her explanation on the graphical representation of parallel lines—opening an additional "window" on the event.

The examples we gave and most of the examples suggested by students were linear, for which three possibilities exist: one solution, no solution, or an infinite number of solutions. Interestingly, in four of the classes a few students came up with several proposals of the following type: the expressions $2x$ and x^2; the solution $x = 2$. Most students did not bother to write an equation to check their answers by using the computer to solve the equation. Only one did, and was surprised. Students and teachers had to deal with the situation: Why is zero a solution? Why did we get two solutions? Thus students' work, CAS performance, and students' reflection became intertwined. Consequently, some students expressed interest in inventing riddles that have more than one solution. One of the teachers challenged the whole class to invent more riddles with two solutions. Students suggested that the equation $1/x = x$ has exactly two solutions, 1 and –1. The teacher asked, "Are you sure?" Some students answered "yes," but others were not sure—even after they had applied the Solve command. The teacher advised them to Plot the graphical representation of the expressions to add conviction. In this situation, the teacher had become a facilitator, using the symbolic and graphical tools of CAS to construct knowledge. The students were quite pleased when the software agreed with their planned solutions. Thus more trust in, and appreciation for, the new environment was established. Occasionally, students expressed uneasiness that the computer simply solves but does not explain how to solve, so the foundation was established for the formal teaching of manipulations on (in)equations.

"Warring Expressions"

The unit "Warring Expressions" was implemented in the six classes after the students had formally learned to solve (in)equations and had worked on modeling word problems (Zehavi and Mann 1999). The game described in the previous section introduced another kind of model. The students in all six classes liked to construct and play the game several times using the software and accumulated the results in class. The question of fairness was raised spontaneously, with some anger expressed over the random results obtained. The students definitely became involved in this game. The teachers asked them to write and explain what they thought about the fairness of the game. A few students said that the game is fair because it is played with random numbers. Others said that on the one hand it is fair, but on the other hand the second expression wins in almost all cases. This dilemma was found in approximately 20 percent of the students' notes. About 40 percent of the notes dealt with the problem arithmetically, with students' expressing their arguments as follows: "The second expression is simplified to $4x + 3$, so you multiply x by a big number and add a number;

in the first expression, $2x - 17$, you multiply by a small number and subtract a number." Some of the students' notes mentioned the fact that for -10, both expressions evaluate to the same number. In only 20 percent of the notes, students interpreted the game in terms of solving an algebraic (in)equation $7x + 3(1 - x) > 2(x + 6) - 29$. The teachers discussed the three approaches in their classes. In the discussion, the notions of chance and probability were raised. On the basis of the accumulated results, some students said that the chance for the second expression to win is 0.75. Two teachers suggested that students find the scores of the two expressions for a game in which 100 numbers are drawn. The students noticed that the second expression almost always scored for less than three-fourths of the numbers. This observation prompted the classes to realize that for 30 out of 41 numbers ($30/41 \approx 0.732$) the second expression wins. This realization called for introducing some basic concepts in probability.

Next the students were asked to explain the unfairness of the game by constructing graphical and algebraic representations (see **fig. 11.4**) and linking the two representations. For example, students were reflecting on the meaning of the evaluation of a numerical true-false statement in terms of points on the graph. Then the classes were split into groups trying to "correct" the game. The teachers observed the work of the groups and gave advice only when asked. The students developed two main strategies: (1) modifying the domain of the random numbers, and (2) modifying the expression(s.) **Figure 11.6** graphically illustrates two fair games. On the left-hand side, the domain was extended to numbers from -40 to 20 by changing the random drawing command to r:= -40 + RANDOM(61.) On the right-hand side, the number 20 was added to first expression, which then became $2x + 3$; thus the graph was translated vertically by 20.

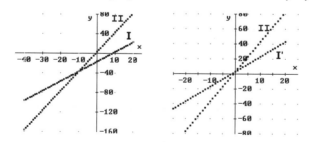

Fig. 11.6. Two modified fair games

The students incorporated a variety of techniques and arguments while working on the tasks. By observing students' work, teachers were able to actually see how students struggled with the mathematical knowledge, how students learned, and what additional help students needed to produce modified games.

Teachers' Workshops

The six teachers met with the team members for one-day workshops three times during the school year. In these workshops, the teachers reported and discussed their findings regarding the units they had integrated into their teaching. They commented on the units and suggested new tasks. The workshops were held in a computer lab. Examples of students' work and teachers' suggestions were demonstrated on the computers. The students' interest in riddles with two solutions led to the cooperative design of a sequence of tasks by the teachers and the project team. The sequence starts by asking students to find two numbers for which the expressions $x^2 + 4x$ and $2x + 3$ are evaluated to obtain the same results. The solutions for this riddle, 1 and –3, can be found by trial substitutions or by writing and solving an equation. In the following task, one expression is $x^2 + 4x$, as before. One number is missing in the second expression, $2x +$ ___. The students are asked to fill in the blank and to find two numbers that solve the riddle (see **fig. 11.7**). This task calls on arithmetic flexibility, as well as algebraic and graphical arguments. It calls for exploration of the possible answer. We can find for which number(s), when the blank space is filled in, the equation will have one solution or no solution at all. To avoid complex solutions given by the software, the students should specify "solve for real." The teachers said that if we introduce riddles with two solutions, students would push for three-solution riddles, so hands-on attempts in this direction were made during the workshop.

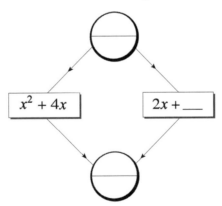

Fig. 11.7. *Complete the riddle, and find two numbers.*

In the workshop that was devoted to the Warring Expressions game, we were interested in hearing details from teachers' observations of students' work. However, the teachers were eager to delve into the probability aspects involved in the game. We had to bargain with them by saying, "You tell us what we need to know to revise the material, and we will teach you how to use the software to design virtual experiments." We upheld our part of the bargain by developing a virtual experiment program at the workshop (see **fig. 11.8**). The domain of drawn numbers is from −20 to 20. The game is played for 1000 numbers. The teachers ran the program several times and were pleased to see that the frequencies were very close to the probability values of the three possible events: 0.732, 0.244, and 0.024.

```
r := -20 + RANDOM(41)
```
A LIST consists of 1000 random numbers from -20 to 20.
```
LIST := VECTOR(r, n, 1000)
              LIST = [12,-3,0,5,-19,8,................]
```
LIST1 and LIST2 are obtained by substitution of the drawn numbers in the expressions.
```
LIST1 := VECTOR(2·(x + 6) - 29, x, LIST)
LIST2 := VECTOR(7·x + 3·(1 - x), x, LIST)
```
We introduce a variable d that denotes the difference between respective elements of the lists.
The command DIMENSION counts the number of the SELECTed values of d.
```
           DIMENSION(SELECT(d > 0, d, LIST2 - LIST1)) = 739
           DIMENSION(SELECT(d < 0, d, LIST2 - LIST1)) = 235
           DIMENSION(SELECT(d = 0, d, LIST2 - LIST1)) = 26
```

Fig. 11.8. A program for virtual games

Next we redirected the discussion to the teachers' observations in the school. They all said that the probability view played a dominant role in students' arguments. Relying on a variety of ideas from their classroom experiences, the teachers suggested a "practically fair" game in which the domain of random integers was increased to [−100 – 100]; in this domain the probability that the second expression gives a larger result is 0.547. We modified the program above to play this game. By a "practically fair" game, the teachers meant a game in which one of the expressions has a small advantage over the other but when played for ten numbers, both expressions can win. In this version, students playing the game can experience the suspense of playing with random numbers without being distracted by the unfairness effect.

The teachers reported that the graphical representation was clear to the students and thus could be useful in helping students understand the algebraic aspects. For example, when the graphs of the two expressions intersect at a point $(0, b)$, the solution of the equation constructed of the two expressions is 0. For linear equations, this outcome means that the two expressions are simplified to $ax + b$ and $cx + b$. The teachers' reaction to the algebraic representation that incorporated the true-false values (see **fig. 11.4**) was that it was intriguing to the students. However, the notation confused some students. These students remembered that when they applied the Solve command to an (in)equation, for which the truth set is the universal set, the output was "true," so in the current example they said, "If true, why false?" The teachers explained to the students that the software uses the same notation when it simplifies a true numerical statement.

Toward the end of the workshop we challenged the teachers to design more almost-fair games. A group of workshop participants devised an advanced game that has three linear expressions and that can be implemented at the end of eighth grade and in the upper grades. Others, influenced by previous riddles with quadratic equations, suggested an almost-fair game using a linear and a quadratic equation (see **fig. 11.9**). In this version, the probability that the linear expression gives a larger result is $23/41 \approx 0.561$.

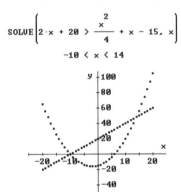

Fig. 11.9. A new "warring expressions" game

The rationale for this game is twofold: (1) design an almost-fair game that will not distract the students' attention toward probabilistic arguments, and (2) make the expressions complex enough that the students will be implicitly motivated to analyze the game with algebraic tools, with less computational preference. We decided to include this new game as an assignment project question in the eighth-grade workbook.

Conclusion

Researchers who focus on students' difficulties in algebra have suggested that to help students overcome conceptual difficulties, we need to design tasks that are intended to bridge the arithmetic-symbolic gap (see, e.g., Wheeler and Lee 1986). The CAS environment enabled us to design tasks involving random numbers and games that lead students to view the two sides of an equation as one entity. Students were motivated to invent riddles with two expressions and to play the Warring Expressions game in the computer lab. This new technological environment enabled them to express themselves at various arithmetic-algebraic levels. In addition, the students could reflect on their work and take more responsibility for their learning. However, in the formative development process we realized that to achieve the main objective, that of helping the student internalize the algebraic structure of (in)equations, we had to improve the sequence of tasks by including nonlinear examples. The arithmetic methods that were helpful with linear equations became less effective with quadratic equations; thus the role of the Solve command was more marked. Initially, we planned to deal with quadratic equations at a later stage of eighth grade, more specifically, using the Solve command as a "black box" in modeling problems. As a result of the feedback from the experimental version of the units, we modified our original plans.

In introducing graphical representation in the analysis of the riddles and the games, we wanted to take advantage of a functional approach to algebra, an approach that helps students visualize two expressions as parts of one (in)equation. In practice, viewing the graphs on the screen was not enough; we had to design tasks in which the students were explicitly instructed to link graphical representations with the arithmetic-algebraic representations that were created using the software (see **fig. 11.4**). On the one hand, such tasks helped to elicit some concise symbolic representation. On the other hand, the syntax and notation that the software used confused students. In performing these new tasks, the students became familiar with a variety of CAS symbolic, graphic, and programming features that became, through experience, an integral part of their mathematics. However, we should be aware that talking in the language of the software could be problematic, especially for students just beginning algebra.

A natural link to probability was created because of the unfairness of the Warring Expressions game. At teachers' request, we constructed virtual experiments for the game by using the random generator and other functions of the software. Teachers suggested designing similar tasks for explaining elementary concepts of probability, which are included in the junior high school syllabus.

The task-design process described in this chapter had ramifications with respect to the teacher-training program. To be able to effectively guide students in using the various tools of the software interactively, teachers first need to review the relevant mathematical methods for solving equations in the real and complex field. They also need some exposure to learning events that have the potential to intertwine student work, CAS performance, and student reflection.

Acknowledgments

The authors are grateful to Carolyn Kieran for her constructive comments regarding a previous CAS-based study of the MathComp project and to Lea Wasserteil for her contribution to the development of the learning units that are discussed in this chapter.

References

Aspetsberger, Klaus, Karl Fuchs, and Fritz Schweiger. "Fundamental Ideas and Symbolic Algebra." In *The State of Computer Algebra in Mathematics Education*, edited by J. Berry and J. Monaghan, pp. 45–51. Bromeley: Chartwell-Bratt, 1997.

Guin, Dominique, and Luc Trouche. "The Complex Process of Converting Tools into Mathematical Instruments: The Case of Calculators." *International Journal of Computers for Mathematical Learning* 3, no. 3 (1999): 195–227.

Harper, David, Chris Wooff, and David Hodgkinson. *A Guide to Computer Algebra Systems*. Chichester: John Wiley & Sons, 1991.

Heid, M. Kathleen. "Resequencing Skills and Concepts in Applied Calculus Using the Computer as a Tool." *Journal for Research in Mathematics Education* 19, no. 1 (January 1988): 3–25.

Judson, Phoebe T. "Calculus I with Computer Algebra." *Journal of Computers in Mathematics and Science Teaching* 9, no. 3 (1990): 87–93.

MathComp Team. *Computers and Mathematics Education—a Textbook for Teachers*. Rehovot, Israel: The Weizamnn Institute of Science, 1998.

———. *Derive Workbooks for Grades 7–9*. Rehovot, Israel: The Weizmann Institute of Science, 1999.

Noss, R., and C. Hoyles. *Windows on Mathematical Meaning: Learning Cultures and Computers*. Dordrecht, The Netherlands: Kluwer Academic Publishers, 1996.

Palmiter, Jeanette R. "Effects of Computer Algebra Systems on Concept and Skill Acquisition." *Journal for Research in Mathematics Education* 22, no. 2 (March 1991): 151–56.

Pozzi, Stefano. "Algebraic Reasoning and Computer Algebra Systems: Freeing Students from Syntax?" In *Derive in Education: Opportunities and Strategies*, edited by H. Heugl and B. Kutzler, pp. 171–88. Bromley: Chartwell-Bratt, 1994.

Small, Don B., John M. Hosack, and Kenneth Lane. "Computer Algebra in Undergraduate Instruction." *The College Mathematics Journal* 17 (1986): 423–33.

Tall, D. "Computer Environments for the Learning of Mathematics." In *Didactics of Mathematics as a Scientific Discipline*, edited by R. Biehler, R.W. Scholz, R. Strässer, and B. Winkelmann, pp. 189–99. Dordrecht: Kluwer, 1994.

Wheeler, D., and Lesley Lee. "Towards a Psychology of Algebra." In *Proceedings of the Eighth Annual Meeting of the North American Chapter of the International Group for the Psychology of Mathematics Education*, edited by Glenda Lappan and Ruhama Even, pp. 133–38. East Lansing, Michigan: Michigan State University, 1986.

Zehavi, Nurit. "Challenging Teachers to Create Mathematical Projects with Derive." *The International Derive Journal* 3, no. 2 (1996a): 1–16.

———. "Establishing Relationships between Formal Notations and Procedures for Performing Mathematical Tasks." In *Proceedings of the International Derive and TI-92 Conference*, edited by B. Barzel, pp. 543–58. Munster: Wilhelms University, 1996b.

———. "Changes That Computer Algebra Systems Bring to Teacher Professional Development." In *Proceedings of the Twenty-first Conference of the IGPME*, edited by E. Pehkonen, pp. 307–14. Lahti: University of Helsinki, 1997a.

———. "Diagnostic Learning Activities Using Derive." *Journal of Computers in Mathematics and Science Teaching* 16, no. 1 (1997b): 37–59.

Zehavi, Nurit, and Giora Mann. "The Expressive Power Provided by a Solving Tool: How Long Did Diophantus Live?" *The International Journal for Computer Algebra in Mathematics Education* 6, no. 4 (1999): 249–65.

Activity 6

The three numbers a, h and b form a *harmonic sequence* if their reciprocals,

$$\frac{1}{a}, \frac{1}{h} \text{ and } \frac{1}{b},$$

form an arithmetic sequence. The number h is called the harmonic mean of a and b.

(a) Find the harmonic mean of $\frac{1}{6}$ and $\frac{1}{12}$.

(b) Express h in terms of a and b.

(c) Show that $\frac{1}{a}, \frac{1}{h}$ and $\frac{1}{b}$, form an arithmetic sequence.

Solution:

- solve $\left(\dfrac{1}{h} - 6 = 12 - \dfrac{1}{h}, h\right)$ $\qquad$ $h = \dfrac{1}{9}$

- solve $\left(\dfrac{1}{h} - \dfrac{1}{a} = \dfrac{1}{b} - \dfrac{1}{h}, h\right)$ $\qquad$ $h = \dfrac{2 \cdot a \cdot b}{a+b}$

- $\dfrac{1}{h} - \dfrac{1}{a} \Big| h = \dfrac{2 \cdot a \cdot b}{a+b}$ $\qquad$ $\dfrac{a-b}{2 \cdot a \cdot b}$

- $\dfrac{1}{b} - \dfrac{1}{h} \Big| h = \dfrac{2 \cdot a \cdot b}{a+b}$ $\qquad$ $\dfrac{a-b}{2 \cdot a \cdot b}$

or

- comDenom $\left(\dfrac{1}{a} + (n-1) \cdot \left(\dfrac{1}{\frac{2 \cdot a \cdot b}{a+b}} - \dfrac{1}{a}\right) \Big| n = \{1\ 2\ 3\}\right)$

$$\left\{\dfrac{1}{a}\ \ \dfrac{a+b}{2 \cdot a \cdot b}\ \ \dfrac{1}{b}\right\}$$

Activity 6

Lines 1 and 2: Use the fact that in an arithmetic sequence each term differs from the one before it by the same amount, $d = a_n - a_{n-1}$.

Lines 3 and 4: Check that the differences are the same.

Line 5: As an alternative form of the check, use the formula of the nth term of an arithmetic sequence, $a_n = a_1 + (n-1)d$, with $d = \dfrac{1}{h} - \dfrac{1}{a}$ to find the first three terms ($n = 1, 2,$ and 3).

PART III

Evidence and Implications from Research: Introduction

THIS SECTION describes the evidence that research offers teachers, curriculum developers, and researchers in using computer algebra systems to enhance students' mathematical understanding. Individual chapters describe several areas of research. The authors foster insight into students' thinking and describe ways that CAS use can encourage such thinking among students in the seventh through twelfth grades. Descriptions of empirical studies also include interesting CAS-relevant tasks that can be used immediately in secondary school classrooms.

Zbiek's initial overview of research evidence contains a set of explicit implications from a broad collection of studies covering many aspects of CAS use. The topics include attitude change, problem-solving success, development of manipulation skills, use of graphical representations, and symbolic reasoning. Comparisons of students using CAS with students not using the technology are also included. This overview of research provides a rationale for using CAS and makes suggestions for effective teaching in a CAS-equipped classroom. Each of the following three chapters discusses a focused area of research. Each chapter contains rich examples of CAS-relevant tasks and samples of students' thinking.

Cedillo and Kieran describe successful experiences of eighth-grade students with access to computer algebra systems in introductory algebra settings. In "Examples of the Activities" they include CAS tasks that helped early algebra learners move beyond their arithmetic, prealgebraic understandings. Excellent examples of follow-up questions are included for teachers as they help their students better understand such important ideas as algebraic equivalence. Cedillo and Kieran's review of research-related literature clearly links this CAS-based approach with students' learning of language.

Using examples from ninth-grade students' algebra work, Drijvers describes the close relationship among paper-and-pencil work, computer algebra, and mathematical thought. In this research, the activities used with students help them understand essential issues, such as algebraic equivalence. Drijvers' elegant descriptions of instrumentation and process-object understanding and the ways they relate to CAS use are valuable tools for thinking about learning and teaching mathematics in technological settings.

Lagrange reminds educators that learning skills need not be equated with learning mindless manipulations. He presents a unique way of thinking about the interplay of concepts and techniques in secondary school students' understanding of mathematics. His analysis of calculus and precalculus tasks parallels the approach that Drijvers uses with less mathematically mature learners. This chapter paints a vivid picture of what might be expected when students use CAS in learning and doing mathematics. Focused interpretations of particular studies present evidence about ways to design and conduct CAS-related lessons.

—Rose Mary Zbiek

CHAPTER 12

Using Research to Influence Teaching and Learning with Computer Algebra Systems

Rose Mary Zbiek

EFFECTIVE USE of computer algebra systems to enhance mathematics learning, teaching, and curricula requires thoughtful examination of related research. Therefore, this chapter begins with a summary of the *nature* of research and follows with a discussion of *existing* research and its implicit messages for educators who wish to implement practices involving computer algebra systems (CAS) in their classrooms or to conduct future research studies. For convenience, "bottom line messages" for mathematics teachers, supervisors, and curriculum coordinators appear as framed inserts below the paragraph(s) they summarize in the "Trends and Implications" section of this chapter.

Research Reports, Access and Availability

During the last two decades, computer algebra systems have entered mathematics classrooms at varying rates in many countries. In general, CAS-related research has lagged behind the implementation of CAS projects. As a result, the emerging pool of research is minimal and fractured. Authors not uncommonly write and publish articles that involve CAS classroom activities rather than focus on carefully conducted empirical studies. This chapter draws heavily from articles that report actual research studies but also makes reference, when appropriate, to other forms of scholarship. Citing papers in the latter group helps to reinforce incidents reported in formal research studies and to identify pressing issues in the classroom Conference papers, book chapters and doctoral dissertations will be used to illuminate those things that students do, learn, and experience in classrooms in which CAS are used. Many of the papers cited as sources in this chapter originally appeared in professional mathematics publications and research journals or were presented at conference proceedings addressing a variety of specific topics, including the use of commercial CAS products in the classroom.

Research Studies, Context and Characteristics

As educators reflect on the results of specific research studies, they should consider the context and characteristics of the investigations; educators should also consider how the studies might influence teaching and learning. The mathematical and theoretical contexts of CAS-related studies affect the outcomes that researchers write about and the ways their papers are used to influence classroom practice.

Mathematics context. Literature related to using CAS covers a variety of topics, from early algebra to differential equations. The literature also covers topics taught to prospective mathematics teachers or associated with the intense problem solving taught to students in their first two years of college. The majority of CAS-related research addresses upper secondary and college students enrolled in calculus and precalculus courses. However, the current discussion focuses on secondary school studies, augmented with research conducted in college mathematics, engineering, and preservice teacher education classes. All references to specific studies include a brief description of the students who participated in the studies and an overview of the mathematics content included in the studies. Because an understanding of CAS at the secondary level requires a consideration of international work, chapters by Cedillo and Kieran, Drijvers, and Lagrange appear in this section. Such countries as France, Austria, and Australia have national expectations regarding the use of CAS in secondary schools and have tied the implementation of CAS projects to extensive studies of CAS potential; for example, Austria has acquired a national site license for Derive.

CAS-related literature focuses on many different computer algebra systems. Recent studies frequently use Texas Instruments TI-92 and TI-89 calculators, along with such programs as Derive or *Mathematica*. Research studies conducted in the 1980s and early 1990s often focused on Calculus T/L II, Mathematics Exploration Toolkit, or MuMath. The notable differences among these products are user interfaces, syntax rules, portability, the computer systems on which they operate, and the public or private nature of their screens. When possible, a reference to a particular study includes the name of the computer algebra system(s) used in the study.

Research studies that focus on using CAS in the classroom are seldom brief. Most investigators use combinations of treatments and data collection that span a period of several months. Many research studies center on students who use CAS for at least one month and often for as long as one semester or an entire school year. Existing research suggests that the initial experience of teaching with CAS can be extremely challenging for teachers, and teachers' instructional strategies unfold and change in unexpected ways during teachers' early attempts in implementing CAS in their classrooms. All studies should take into consideration the time required for teachers to adjust to the technology. Researchers should also be cognizant of the changes in curriculum and assessment practices that frequently accompany the introduction of CAS and should include information that states whether the research was conducted in the classrooms of teachers who were new to teaching with this form of technology.

Three factors inhibit brevity in describing a CAS-related research study: (1) the duration of the studies, (2) the accompanying complexity of data sources, and (3) the breadth of the mathematics addressed. The references to CAS-related studies include details as needed and note whether a specific research treatment or condition involves essential features, such as CAS-use projects, collaborative learning, or the mathematical modeling approach. The fact that these features affect the results should be obvious to the researcher and of less concern to the classroom teacher than the extent to which they affect how the results influence decisions in classroom.

Theoretical context. Some of the lack of coherence in CAS-related literature may be attributed to the variety of perspectives and interests of the authors who write about using CAS. In the United States, studies involving CAS were often designed to test or evaluate an alternative curriculum. For example, several papers dealt with the test-trials of the *Calculus and Mathematica* curriculum materials. Other writers focused on CAS within the context of cognitive actions and psychological constructs. Their individual areas of interest included foci as different as reification of function concept, acquisition and application of problem-solving heuristics, and gender differences. If the theoretical framework within which a study was conducted significantly affected the interpretation of the results, theoretical context is mentioned.

Research reports. The different perspectives of authors not only affect the research but also influence how it is reported. One drawback in most published research is the absence of information that explains exactly how students and teachers used CAS in classroom activities and whether students had access to CAS outside the classroom. Such omissions may indicate that the main focus of the researchers was to develop theoretical frameworks or to test hypotheses for a general view of teaching and learning rather than to study the effects of the technology in the classroom. For example, Galindo's (1995) interest in visualization is reflected in a descriptive explanation of the details surrounding visualization issues but contains virtually no details about the mathematics tasks and CAS activities used in the study. Although research reports may lack important information about technology use by students and teachers, their results—positive or negative—may be used to support or disparage CAS use.

Unfortunately, information that is left out of research reports is an impediment to educators who are looking for strong and promising examples of CAS activities for the classroom. In fairness to the authors, the details of the mathematics and technology use may be too extensive to include in the usual confines of printed publications. Sufficient space may not be available—particularly in conference proceedings—to accommodate complex mathematics problems, complete descriptions of student work, and displays of technology results. As a result, authors may be unable to include descriptions of open-ended problems, long-term projects, and student-led explorations. Therefore, many reports are not extremely useful to teachers as they develop CAS-rich curriculum materials or decide how to teach with CAS. However, available research does echo several themes that may influence the decision to use or not use CAS. Research can illuminate general expectations and cautions for teaching and learning with CAS. Subsequent chapters in this section provide rich examples of students' use of CAS and mathematical thinking.

Trends and Implications

Despite the variations among chapters, several trends emerge that may partially answer the question *"What messages does research provide regarding the use of computer algebra systems in teaching and learning secondary school mathematics?"*

Effects on attitude. The use of CAS does not automatically lead to positive changes in students' attitude, according to obvious trends emerging from studies of students of various ages and mathematics levels. Students' attitudes typically became slightly less positive or remained basically unchanged during one semester or more of using CAS. However, a study conducted by Hopkins and Kinard (1998) on the attitudes of college developmental algebra students using CAS resulted in an exception to the "rule" above. The researchers reported a statistically significant improvement in student attitude in a TI-92 class but observed no improvement in attitude among students in the other classes, the control group. The CAS-using students also had better attitudes in the mathematics course that followed and passed the subsequent course at a remarkably higher rate than the students who had not used CAS in the preceding course. One is tempted to attribute the favorable changes in students' attitudes to their use of CAS. However, most of the students reported believing that the three factors most influential to their learning were, in order, (1) the teacher, (2) the curriculum, and (3) the CAS. The importance of "teacher over tools" parallels the finding of O'Callaghan (1998) in his studies with college algebra students. The only attitudinal difference between his CAS-using group and the control group was a more positive rating for the researcher as teacher, a score given by students in the CAS group. Of course, the students in his study may have attributed their attitude changes to the teacher when they should have attributed the changes to other contributing factors.

> *In general, students who begin their CAS experience with a positive or less skeptical view of learning mathematics with technology may eventually conclude that technology is not a panacea. In our students' opinions, we, their teachers, continue to be the most significant factor in maintaining or improving their attitudes.*

Evidence exists to suggest that novice learners or previously unsuccessful mathematics students may be more open to nontraditional teaching methods associated with CAS than experienced mathematics students, who may resist the innovation. Yet, novice students often struggle with curriculum materials that lack in-depth coverage of information new to them and that contain what they perceived to be inadequate description of manipulations or too few examples. Using Derive with precalculus, calculus, and linear algebra topics, Pierce and Stacey (in press) concluded that the more advanced students who studied these topics attributed their increased "math confidence" to their use of CAS.

The students also stated that curriculum materials, such as worksheets and the use of multiple representations, contributed to their understanding. In a separate report of the study mentioned above, Pierce and Stacey (forthcoming) noted that students involved in the study believed that the use of CAS enhanced their confidence but thought that they actually learned more from paper-and-pencil experiences. The relationship among student confidence, CAS use, and skill acquisition differed among individual students, as was expected. Bergsten's (1996) Swedish engineering students indicated that to "learn" analysis required by-hand work. They viewed the issue to be one of "control" and did not perceive that they were in control when the computer algebra system, Maple, executed the symbolic manipulations for them. As a result, they appreciated the computer algebra system as a problem-solving tool but not as a learning tool. Curiously, they liked the graphing aspect of the computer algebra system and did not raise the control issue with respect to CAS graphs.

> *Student attitude and mathematical confidence may be greatly affected by students' perceptions of learning and succeeding in mathematics and by their perceptions of the appropriateness of the curriculum materials used in class. As teachers, we need to consider how well the CAS curriculum that we have selected for use in the classroom helps our students learn and succeed in this new environment.*

Studies of students' attitude may be part of a larger issue—how students relate to CAS and how their attitudes and understanding of mathematics emerge through that relationship. The research on students' attitude suggests that students begin with positive attitudes, in general, and leave with similar attitudes; their original enthusiasm appears to be partially diminished by the realization that using CAS does not totally relieve them of their mathematical learning obligations. To understand the cognitive link between student mathematics and CAS, Lagrange (1999) adapts the idea of "instrumental genesis," as found in Verillon and Rabardel (1995), to the use of CAS. Instrumental genesis conveys the idea that students who are learning an area of mathematics for the first time learn it within the confines of the mathematics of the tool they are using. While students are making sense of the mathematics embodied in the tool, they are simultaneously using the tool to make sense of the mathematics. In chapter 14 in this volume, Paul Drijvers uses instrumentation as a method of reflecting on the ways that paper-and-pencil experiences and mental explorations relate to secondary school students' experiences with computer algebra systems.

> *As students learn new mathematics with CAS, their mathematical understanding will reflect, in significant ways, the mathematics of the CAS they are using. As teachers, we can expect students' CAS-related mathematics knowledge to be very rich, but it may appear to be at least slightly different from our knowledge.*

Comparison studies. Studies that compare classes that use CAS with classes that do not use CAS range from more formal studies to less formal, local data-driven evaluations of particular course experiences. Many research papers revolve around the use of CAS in calculus courses. Using different degrees of research rigor, Alarcon and Stoudt (1997), Lefton and Steinbart (1995), Porzio (1995), and Roddick (1995) evaluated the potential of the *Calculus and Mathematica* curriculum in their respective college mathematics departments. Roddick's (1995) interviews with both *Calculus and Mathematica* students and comparison-course students enrolled in subsequent engineering courses further suggest that students can leave CAS experiences and feel sufficiently comfortable in future science or engineering classes to "attack" application problems with a combination of procedural and conceptual approaches.

Some studies focus on students' learning of algebra. In teaching algebra at the college level, O'Callaghan (1998) used the secondary school curriculum Computer-Intensive Algebra, and Mayes (1992, 1993, 1995, 1998) used his college-intended Applications, Concept, and Technology (ACT) in Algebra curriculum (Mayes and Lesser 1998). Both researchers compared CAS approaches with "traditional" course material. However, Mayes evaluated the curriculum, whereas O'Callaghan addressed the extent to which students could reify the function concept (i.e., reach a point where a function was an object as well as a computational process). Working strictly with secondary school students who were using the *Computer-Intensive Algebra* curriculum, Heid (1992) compared students using the *Computer-Intensive Algebra* curriculum with students in traditional algebra and prealgebra courses.

> *Comparison studies usually involve two or more very different curricula, one that uses CAS to teach certain mathematical ideas and one that, without CAS, emphasizes very different mathematics. Such studies often indicate that the use of CAS—when paired with a curriculum enhanced by mathematical modeling—can lead to deeper conceptual understanding and better application performance, accompanied by similar or slightly depressed computational performance. These results often take the form of better CAS-group performance on a researcher-devised concept test or application test and comparable or lower CAS-group performance on a common, skills-based final examination. Such outcomes serve to remind us, as teachers, that our students tend to learn only what we teach them, in either of two very different settings. Yet skill performance is often very similar in both groups.*

More rare are studies that use identical teaching materials in comparing CAS-use and non-CAS-use classes. Gaulke (1998) studied the use of Derive in *Brief Calculus*, a calculus course of a very introductory nature. Students in the treatment group scored higher than students in the control group—both groups given identical computational posttest items. Gaulke found that treatment-group students were correct more often when they used Derive than when they worked by hand—not an unexpected discovery. Noteworthy is the fact that the students in this study were expected to generate and write the Derive commands on class tests because it was a requirement of the course; this expectation may have contributed to their success in the use of Derive and their success on the conceptual posttest items.

> *We, as teachers, should expect our students to generate and write CAS commands themselves if they are learning computational skills in a classroom in which CAS are available.*

Connors (1995) presents minimal evidence that using CAS has potential gender-equity discrepancies. The computer algebra system Theorist, when used with other technology in a calculus course, resulted in students' earning higher—but not significantly higher—scores on a common final examination than those students enrolled in a traditional course. Females in the CAS group scored higher on the examination despite the fact they entered with lower SAT-Mathematics scores than males and than females in the control group.

> *On the basis of only minimal evidence available at this time, using CAS may have benefits for our female students, benefits that are similar to those found with the graphing calculator, as summarized in Dunham and Dick (1994).*

Most of the comparison papers focus on what was done and what should be done to improve mathematics curriculum and teaching. Importantly, reports of failed attempts at using CAS in mathematics education (e.g., Alarcon and Stoude [1997]) suggest that the failure can be attributed to the methods behind the preparation and planning for the changes associated with using CAS rather than to inherent flaws in the concept of using CAS in the classroom. Comparative studies are less common in countries with national curricula and often focus on ways that students and teachers learn and on what they do with technology. In chapter 15 in this volume, Lagrange reflects the latter perspective in the context of learning techniques.

Problem solving. Another subset of CAS research involves studies that target problem solving. For example, Matras (1988), Santos (2000), and Runde (1997) focused on problem solving and used CAS as a way of varying heuristics used in solving standard algebra word problems. A concentration on this narrow aspect of mathematics content may drastically alter the findings and lead to students' development of a very narrow set of CAS uses and skills. Matras (1988) uses her results obtained from a study with high school algebra students using MuMath to illustrate that students who use CAS along with alternative curricula may learn and implement certain problem-solving strategies better than their peers who do not use CAS. Conversely, evidence suggests that students may not actually learn the intended problem-solving procedures. Runde (1997) noted that using the TI-92 calculator allowed college students in basic algebra courses to try more equations to model a word problem; students who did not use CAS could try only one equation. As a result, the CAS may have facilitated students' guess-and-test strategies and hampered their acquisition of essential equation-writing skills needed to solve standard word problems efficiently. Of course, the inconsistencies in the results may reflect differences in the individual curricula used during the study.

> *Student use of CAS, in combination with appropriate course materials, can influence—positively or negatively—how students solve problems. We, as teachers, need to emphasize more than final answers and monitor our students' methods of using CAS.*

Skill issues. Research related to the use of CAS as it affects skill acquisition suggests that using CAS does not harm students and may increase the speed and ease with which they learn about symbols and the relationships among symbolic forms. For example, Brown (1998) used TI-92 calculators with Australian junior high school students. The students looked at patterns in CAS results to see symbolic forms and to study factoring; they also explored the "vertex form" of a quadratic expression. At least 60 percent of the students assessed were able to write factorable expressions and read quadratic factors from a graph. Most of the students could read points on the graph with a vertex to write the equation, but few chose to use the vertex form of the equation in completing the process. Brown concluded that using CAS was not detrimental to the symbolic performance of these students, ages 13–16.

Heid (1992) reported similar results with using CAS in a classroom in which by-hand skills were not explicitly taught until the last three to five weeks of a first-year high school algebra course. The students in her study used the *Computer-Intensive Algebra* curriculum and the Mathematics Exploration Toolkit. The CAS-using students were ability grouped between the high-achieving students who took algebra in grade 8 and those students who began algebra in grade 9. Despite the delayed, minimal emphasis on skills, the CAS-using group performed "between" the other two groups on a common, skill-based final examination. This outcome suggests that their skill acquisition was not harmed by the CAS experience. These findings with secondary school algebra students are similar to those with college calculus students using MuMath in an earlier study (Heid 1984).

> *Students may actually learn skills faster if we, as teachers, first teach them algebraic ideas from a conceptual approach using CAS, with an emphasis on function and mathematical modeling.*

Evidence suggests that using CAS may help students who cannot show mastery of essential skills from previous mathematics courses. Shaw, Jean, and Peck (1997) used Derive with an intermediate algebra course for college students in the developmental mathematics program. They found no significant difference in grades earned in a subsequent statistics course between the students who had previously used CAS in a special developmental mathematics course and the students who had not been selected for this class.

> *Encouraging our students who lack manipulation skills to use CAS may enhance rather than impede their success in future mathematics courses.*

Another fundamental aspect of skill fluency is choosing the correct method at the right time. Using CAS may help with this important component of skill fluency, according to Keller and Russell (1997). Their results showed that college calculus students who used the TI-92 symbolic calculator chose a correct method on a common departmental final examination more often than their peers who were enrolled in CAS-free calculus classes. When the CAS-using students and their non-CAS-using peers chose correct methods, the CAS-using students were more likely also to arrive at a correct solution—using the TI-92 on the examination—than were their non-CAS peers. The results of a study by McCrae, Asp, and Kendal (1999) seem to blend this idea of the improved strategy choice for solving calculus problems found by Keller and Russell with the improved performance by low-ability students found by Shaw and colleagues. McCrae's team followed year-12 Australian students using TI-92 calculators through a unit on differentiation. Although the students were deemed to be in the lowest ability group in their setting, these students' by-hand skills did not suffer on a multiple-choice examination. In addition, the students seemed to know that they did not necessarily have to use the CAS on questions for which its use would not be helpful.

> *Our students, at least those in upper level mathematics courses, can become savvy CAS users and employ the tool to enhance their mathematical performance.*

Some sign of interest has also been shown in CAS-like programs or alternative user interfaces that give learners access to a subset of the usual CAS commands. Strickland and Al-Jumeily (1999) report trials of Treefrog in England. The Treefrog program has less CAS functionality than most CAS. It is designed with a simple user interface and serves as a tool to determine whether students' polynomial manipulations are correct. Students ages 12 to 14 used Treefrog to perform manipulations involving parentheses, such manipulations as expanding the product of a constant and a polynomial expression with the distributive property, along with other manipulations, such as multiplying two binomial expressions. The control group participated in identical lessons but used only paper and pencil to perform computations. The intervention lasted for only 250 minutes and resulted in significantly higher scores for the control group than for the treatment group, although the scores associated with the treatment group were still very low. In their conclusions, Strickland and Al-Jumeily recommended a mixed diet of limited tools, such as Treefrog, along with traditional CAS. This approach contrasts with that presented in this volume's chapter 13, by Cedillo and Kieran, who provided students of similar ages with TI-92 calculators—full computer algebra systems—but used activities that involved only a small portion of the CAS command set. Their results favor the CAS group. Working with the TI-92 and secondary school students ages 15–17, Guin and Trouche (1999) also recommended giving students a limited number of commands.

> *We, as teachers, should feel comfortable allowing our students who are younger or less mathematically sophisticated to use CAS tools, as long as the mathematical tasks are appropriate for the CAS and for the learners.*

Believing that by-hand work is crucial to conceptual development, Monaghan (1995), in a review of CAS use in secondary school calculus, addresses the issue of the manipulations that need to be done by hand. Skarke and Koenig's (1998) technical college teachers raised the same question. As McCrae and colleagues (1999) noted, a potential paradox exists because strong algebraic facility may be needed to obtain significant benefits from CAS work. The existing literature includes several existence proofs that students in CAS-using courses can learn skills well if skill development is delayed. However, the literature sheds less light on the issue of the degree to which each skill needs to be performed by hand for students to build a solid conceptual basis for understanding. Although difficult to investigate, the relationship among by-hand skill acquisition, conceptual development, and the use of CAS in various areas of mathematics—not only in calculus—should be a priority in future research. Drijvers, in chapter 14, and Lagrange, in chapter 15 of this volume, outline important ways of using CAS that transcend mere mastery of routine manipulations and lead to rich understanding of skills.

Graphical emphasis. Graphical representations seem to abound in CAS-using classrooms. Some of the studies involve translations among representations (e.g., O'Callaghan [1998]; Porzio [1995]). These two and other studies report that CAS-using groups were more successful than non-CAS groups with translation items on posttests. However, in all instances, the CAS groups and the comparison groups used different curriculum materials. The positive results of the CAS-using groups may be due largely to the success of the curriculum materials in capitalizing on the technology. Other writers (Mayes 1993) address graphical methods as alternatives to symbolic solutions to certain types of mathematical problems. Some researchers and the students in their studies emphasized graphical interpretation in much less obvious, perhaps inadvertent, ways. Bergsten (1996) reported that his engineering students valued Maple more for its graph-generation capability than for its symbolic manipulation. Heid and her colleagues (1998) did not dwell on graphing, but the sample student work that they described emphasized graphical representations and placed less emphasis on symbolic forms. All these writers—or at least their students—valued the graphical methods available, although not all writers explicitly addressed the possibility that graphs, rather than symbolic forms, could be the "delight" that users find in their CAS capabilities.

Curiously, CAS-focused research reports included many nonsymbolic graphing tasks in discussions of the mathematics that students experienced in classes, on tests, or during interviews. A few writers (e.g., Hillel et al. [1992]) use only graphing tasks to describe what students did in the CAS environments. Graphing tasks also dominated the TI-92 uses reported by Austrian technical college teachers in Skarke and Koenig (1998). Not always clear is whether the symbols and graphs are treated as two comparably important representations or whether the attention given to graphing is so heavy that it skews the intent of the research. This graphing presence raises a question as to whether some of the research studies that seem to be CAS studies are actually graphing-utility studies. In some situations, the positive or negative results may be so highly charged from the intense use of graphical representations that the research does little to influence the use of CAS or capitalize on its symbolic component.

Alternatively, the outcomes could represent additional strength of conceptual understanding when graphs and symbols are used in concert and when symbol manipulation, as well as graph construction, is allocated to the tool. In fact, a blend of graphical and nongraphical methods in the classroom may be very important to students who possess certain characteristics. For example, Nasser and Abou-Zour (1997) saw a significant difference in performance on calculus and precalculus tasks among college students using Derive. Interestingly, the researchers found that students with differences in field dependence did not automatically gravitate toward or away from the CAS. However, students with higher or lower field-dependence scores were more or less successful when using Derive. Nasser and Abou-Zour attributed this result to the extent to which the students used graphs in their CAS solutions. Similarly, Galindo (1995) studied the visual preferences of college students in three types of calculus courses: (1) a class using graphing calculators, (2) a class using *Mathematica*, and (3) a class using no technology. He found that "visualizers" preferred graphical methods, whereas "nonvisualizers" preferred symbolic approaches on mathematical tasks. The results also showed that the nonvisualizers performed significantly better than visualizers in both the nontechnology section and in the *Mathematica* section. Galindo attributed the lack of success for visualizers in the *Mathematica* section to their struggle with the CAS syntax, thus implicitly suggesting that the user interface of CAS is an important consideration in the classroom. However, since each of the three calculus sections used a very different textbook from the others, we cannot know the extent to which the technology alone influenced students' preference.

Instructors' preference for graphical representations constitutes a second major issue. Kendal and Stacey (1999, 2000, forthcoming [b]) studied two upper secondary school teachers in Australia as they taught lessons on differentiation and integration. One teacher clearly preferred graphical forms, whereas the other emphasized symbolic forms. The researchers described the teachers as "privileging" one representational form or another. The reasons that the teachers "privileged" different representations are complex; their choices stemmed in part from individual differences in their definitions of mathematics. The teachers' privileging of different representations in the classroom was apparent in their students' posttest performance.

> *Interestingly, many studies of the use of CAS depend highly on the graphing component of the tool. The graphing work seems to be especially prevalent in situations in which mathematical modeling or mathematical applications to other fields is given a high priority. Having access to graphs may also assist students in developing conceptually richer understandings. Our explicit and implicit choices of CAS, of curriculum material, and of representations in the classroom may either elevate or undermine efforts to implement effective use of CAS in mathematics teaching.*

Additional investigational studies are needed to unravel the impact of graphical emphasis at the potential expense of symbolic forms; studies are also needed to determine which representational forms are most prominent in the classroom and which representational forms are the most effective tools to enhance students' understanding. Kendal and Stacey (forthcoming [a]) investigate the preceding dilemma through a study of a differentiation unit in upper secondary school. Potential relationships regarding the ways different representations may be used at the beginning algebra level can be found in the work of Cedillo and Kieran in chapter 13 of this volume.

Symbolic reasoning. As a result of the increased attention that graphs and symbolic manipulation have attracted, one can easily overlook CAS's impact on other aspects of symbolic understanding. Heid and her colleagues (1998) suggested that college students with substantial mathematics experience paid little attention to symbolic reasoning. The students referred to surface-level features or examples without delving into the mathematical depth of the symbolic forms. This finding confirms the suspicion that the presence of CAS is not sufficient to elicit mature mathematical reasoning about symbols. We must ask ourselves, "How do we, as educators, move our students beyond the superficial aspects of mathematical symbols?" Perhaps a partial answer comes in Brown's (1998) contention that symbolic meaning can be developed in mathematical modeling activities using CAS. Other promising insights lie in pursuing the thoughtful tasks used in the research described in this volume by Drijvers in chapter 14 and by Lagrange in chapter 15.

Afterthoughts and Speculations

The current body of CAS-related literature has several limitations and implications that merit a few last observations regarding the role that research might play in the future.

Influence of experience with graphing calculators. To place research under the "CAS banner" is problematic if the symbolic representation is used only to generate graphs. The current fascination with graphical representations may be natural at this point in the evolution of CAS in mathematics education. Teachers using CAS and researchers investigating its use (e.g., teachers in Kendal and Stacey [1999]) may have gravitated toward CAS following an interest in graphing calculators. This transition would account for the emphasis on the graphical representations in the goals of the studies and a similar interest exhibited in classroom practices. The path followed by teachers as they move from using graphing calculators to using CAS is not clear. Kendal and Stacey (1999) describe two very different routes adopted by teacher colleagues. Their work suggests several indicators that can be used to gauge the transition from using graphing calculators to using CAS; these indicators include a teacher's past perception of success with graphing calculators and a teacher's personal preference for particular representations. Studies that investigate ways that students move from learning with the graphing calculator and related tools to learning in CAS environments are not well represented in the existing literature.

Function concept. A second phenomenon that may influence the prevalence of graphing in CAS studies is the appearance of the CAS during a time when students' and teachers' understanding of "function" was gaining international attention. The use of graphs to understand function, combined with the added potential to work with symbolic forms on the CAS, may have encouraged the design of tasks and the inclusion of graphical elements in the classroom. Some people also sense that CAS perform the manipulations and, therefore, that manipulations may be de-emphasized. The viewpoint expressed in the preceding sentence could be interpreted to be a mandate to educators to refrain from using symbolic forms more than absolutely necessary. As a result, the emphasis could shift to other representations—with the graphical representation being perceived as more interesting, more robust, and more informative than numeric forms. This viewpoint is in contrast with those of Cedillo and Kieran in their work regarding the role of numerical representations; see chapter 13 in this volume. Additional studies are needed to define the role and balance of graphic, numeric, and symbolic representations in learning in environments where CAS are used.

> *CAS may facilitate the use of multiple representations. Some representations have been found to be more interesting for students at different grade levels and more appropriate for the exploration of individual aspects of functions.*

Educators need to be mindful of representation presence and choice in the classroom, within and beyond the CAS. Studies must be designed to explore the use of CAS, beyond those current investigations that heavily emphasize developing the function concept with multiple representations. Such studies should explore the use of CAS in developing by-hand manipulation skills and also investigate the relationship between skill acquisition and conceptual understanding.

Conclusion

The growing collection of reports of carefully conducted research and other scholarly papers can influence classroom practice and curriculum development. Teaching with CAS requires an awareness of the role of symbolic versus the role of graphical representations. Using CAS in the classroom will probably not result in substantial attitudinal changes on the part of students in general but may be beneficial to female students. However, we can reasonably expect that—compared with what might happen without CAS use, appropriate curriculum, and supporting pedagogy—students' performance on conceptual and application tasks will improve and their performance on computational tasks in the absence of the CAS will be only slightly less robust. Students' by-hand manipulation skills need not automatically suffer when the CAS enters the classroom. Time and resources for teacher preparation and planning are crucial, particularly given the high expectations for expertise in technology, mathematics, and pedagogy (e.g., Lumb, Monaghan, and Mulligan [2000]; Zbiek [1995]).

Tasks that involve finding patterns in CAS outputs meet with at least moderate success (e.g., Cedillo and Kieran [this volume]; Kendal and Stacey [2000]). Parameter-exploration tasks seem productive but tend to emphasize graphs; teachers need to demand that students use symbolic reasoning and manipulations to support conjectures based on graphs. The absence of detail in the research reports causes difficulty in using the reports to develop deeper insights into the kind of CAS-related mathematical experiences that best support student learning. Writers who wish to have greater impact on classroom practice should consider ways to include in their research reports more information about the mathematical tasks and CAS activities used in the classroom. If a synergistic relationship between research and teaching is to enhance learning, writers may need to seek alternative venues in which to publish.

References

Alarcón, Francisco E., and Rebecca A. Stoudt. "The Rise and Fall of a *Mathematica*-Based Calculus Curriculum Reform Movement." *Primus* 7 (March 1997): 73–88.

Bergsten, Christer. "Learning by Computing? The Case of Calculus and a CAS." In *Proceedings of the Seventh Annual International Conference on Technology in Collegiate Mathematics*, edited by L. Lum, pp. 36–40. Reading, Mass.: Addison-Wesley Publishing Co., 1996.

Brown, Roger. "Computer Algebra Systems in the Junior High School." Paper presented at the Third International Derive/TI-92 Conference, Gettysburg, Pa., July 1998.

Connors, Mary Ann. "Achievement and Gender in Computer-Integrated Calculus." *Journal of Women and Minorities in Science and Engineering* 2 (1995): 113–21.

Dunham, Penelope H., and Thomas P. Dick. "Connecting Research to Teaching: Research on Graphing Calculators." *Mathematics Teacher* 87(6) (1994): 440–45.

Galindo, Enrique. "Visualization and Students' Performance in Technology-based Calculus." In *Proceedings of the Seventeenth Annual Meeting of the North America Chapter of the International Group for the Psychology of Mathematics Education*, edited by D. T. Owens, M. K. Reed, and G. M. Millsaps, pp. 321–27. Columbus, Ohio: ERIC Clearinghouse, 1995.

Gaulke, Scott. "Integrating Derive into Calculus Instruction." Paper presented at the Third International Derive/TI-92 Conference, Gettysburg, Pa., July 1998.

Graham, Ted. "Derive: Passing Acquaintance or Friend for Life?" *The International Journal of Computer Algebra in Mathematics Education* 4 (1997): 129–40.

Guin, Dominique, and Luc Trouche. "The Complex Process of Converting Tools into Mathematical Instruments: The Case of Calculators." *International Journal of Computers for Mathematical Learning* 3 (1999): 195–227.

Heid, M. Kathleen. "Resequencing Skills and Concepts in Applied Calculus Using the Computer as a Tool." *Journal for Research in Mathematics Education* 19(1) (January 1984): 3–25.

———. *Final Report: Computer-Intensive Curriculum for Secondary School Algebra*. Final report for NSF project number MDR 8751499. University Park, Pa.: Pennsylvania State University, Department of Curriculum and Instruction, 1992.

Heid, M. Kathleen, Glendon W. Blume, Karen Flanagan, Linda Iseri, and Ken Kerr. "The Impact of CAS on Non-routine Problem Solving by College Mathematics Students." *The International Journal of Computer Algebra in Mathematics Education* 5(4) (1998): 217–49.

Hillel, Joel, L. Lee, C. Laborde, and L. Linchevski. "Basic Functions through the Lens of Computer Algebra Systems." *Journal of Mathematical Behavior* 11(1992): 119–58.

Hopkins, Laurie, and Amelia Kinard. "The Use of the TI-92 Calculator in Developmental Algebra for College Students." Paper presented at the Third International Derive/TI-92 Conference, Gettysburg, Pa., July 1998.

Keller, Brian A., and Chris A. Russell. "Effects of the TI-92 on Calculus Students Solving Symbolic Problems." *The International Journal of Computer Algebra in Mathematics Education* 4 (1997): 77–97.

Kendal, Margaret, and Kaye Stacey. "Varieties of Teacher Privileging for Teaching Calculus with Computer Algebra Systems." *International Journal for Computer Algebra in Mathematics Education* 6(4) (1999): 233–47.

———. "Acquiring the Concept of Derivative: Teaching and Learning with Multiple Representations and CAS." In *Proceedings of the 24th annual conference of the International Group for the Psychology of Mathematics Education*, Hiroshima, Japan, July 2000.

———. "The Impact of Teacher Privileging on Learning Differentiation with Technology." *International Journal of Computers for Mathematical Learning* 6(2) (2001): 143–65.

———. "Tracing Learning of Three Representations with the Differentiation Competency Framework." Unpublished manuscript.

Lagrange, Jean-Baptiste. "Complex Calculators in the Classroom: Theoretical and Practical Reflections on Teaching Pre-calculus." *International Journal of Computers for Mathematical Learning* 4(1) (1999): 51–81.

Lefton, Lew E., and Enid M. Steinbart. "Calculus and Mathematica: An End-User's Point of View." *Primus* 5 (March 1995): 80–96.

Lumb, Stephen, John Monaghan, and Steve Mulligan. "Issues Arising When Teachers Make Extensive Use of Computer Algebra." *The International Journal of Computer Algebra in Mathematics Education* 7(4) (2000): 223–40.

McCrae, Barry, Gary Asp, and Margaret Kendal. "Teaching Calculus with CAS." Paper presented at the Fourth International Conference on Technology in Mathematics Teaching, Plymouth, England, August 1999.

Matras, Mary Ann. "The Effects of Curricula on Students' Ability to Analyze and Solve Problems in Algebra." Unpublished Doctoral Dissertation, University of Maryland, 1988.

Mayes, Robert L. "The Effects of Using Software Tools on Mathematical Problem Solving in Secondary Schools." *School Science and Mathematics* 92 (1992): 243–48.

———. "The Effects of Computer Algebra Systems on Concept and Skill Acquisition in College Algebra." In *Proceedings of the Fourth Annual International Conference on Technology in Collegiate Mathematics*, edited by L. Lum, pp. 339–43. Reading, Mass.: Addison-Wesley Publishing Co., 1993.

———. "The Application of a Computer Algebra System as a Tool in College Algebra." *School Science and Mathematics* 95 (1995): 61–68.

———. "ACT in Algebra: Student Attitude and Belief." *The International Journal of Computer Algebra in Mathematics Education* 5 (1998): 3–14.

Mayes, Robert L., and Lawrence M. Lesser. *ACT in Algebra: Preliminary Edition.* San Francisco, Calif.: WCB McGraw-Hill, 1998.

Monaghan, John. "Procepts, Readiness and Semantics: Cognitive Factors in the Use of Computer Algebra Systems in Mathematics Teaching." In *Proceedings of the Sixth Annual International Conference on Technology in Collegiate Mathematics,* edited by L. Lum, pp. 265–71. Reading, Mass.: Addison-Wesley Publishing Co, 1995.

Nasser, Ramzi, and Saed Abou-Zour. "Effect of Cognitive Style on Algebra Problem Solving Using Computer Aided Approach versus an Analytical Method." *The International Journal of Computer Algebra in Mathematics Education* 4 (1997): 359–70.

O'Callaghan, B. R. "Computer-Intensive Algebra and Students' Conceptual Knowledge of Functions." *Journal for Research in Mathematics Education* 29 (January 1998): 21–40.

Pierce, Robin, and Stacey, Kaye. "Reflections on the Changing Pedagogical Use of Computer Algebra Systems: Assistance for Doing or Learning Mathematics?" *Journal for Computing in Mathematics and Science Teaching,* in press.

Pierce, Robin, and Stacey, Kaye. *Positive Learning Strategies Encouraged by Using Computer Algebra Systems.* Forthcoming.

Porzio, Donald T. "Effects of Differing Technological Approaches on Students' Use of Numerical, Graphical and Symbolic Representations and Their Understanding of Calculus." 1995, pp. 1–8. ERIC Document Reproduction Service No. ED391665.

Roddick, Cheryl Stitt. "How Students Use Their Knowledge of Calculus in an Engineering Mechanics Course." In *Proceedings of the Seventeenth Annual Meeting of the North America Chapter of the International Group for the Psychology of Mathematics Education,* edited by D. T. Owens, M. K. Reed, and G. Millsaps, pp. 134–39. Columbus, Ohio: ERIC Clearinghouse, 1995.

Runde, Dennis C. "The Effect of Using the TI-92 on Basic College Algebra Students' Ability to Solve Word Problems." 1997. ERIC Document Reproduction Service No. ED409046.

Santos, Manuel. "The Use of Representations as a Vehicle to Promote Students' Mathematical Thinking in Problem Solving." *The International Journal of Computer Algebra in Mathematics Education* 7(3) (2000): 193–212.

Shaw, Nomiki, Brian Jean, and Roger Peck. "A Statistical Analysis on the Effectiveness of Using a Computer Algebra System in a Developmental Algebra Course." *Journal of Mathematical Behavior* 16 (1997): 175–80.

Skarke, P., and E. Koenig. "New Perspectives in Teaching Mathematics Due to the Use of the TI-92." Paper presented at the Third International Derive/TI-92 conference, Gettysburg, Pa., July 1998.

Strickland, Paul, and Dhiya Al-Jumeily. "A Computer Algebra System for Improving Student's Manipulation Skills in Algebra." *The International Journal of Computer Algebra in Mathematics Education* 6(1) (1999): 17–24.

Verillon, Pierre, and Pierre Rabardel. "Cognition and Artifacts: A Contribution to the Study of Thought in Relation to Instrumented Activity." *European Journal of Psychology of Education* 10(1) (1995): 77–101.

Zbiek, Rose Mary. "Her Math, Their Math: An In-service Teacher's Growing Understanding of Mathematics and Technology and Her Secondary School Students' Algebra Experience." In D. Owens, M. K. Reed, and G. M. Millsaps (Eds.), *Proceedings of the Seventeenth Annual Meeting, North American Chapter of the International Group for the Psychology of Mathematics Education*, edited by D. Owens, M. K. Reed, and G. M. Millsaps, 1995, pp. 214–20. Columbus, Ohio: ERIC Document Reproduction Service No. ED 389610.

Activity 7

Solve for x, and check: $5x - 2(x-7) = 8(x+2) - 17$. Show your work by not using the built-in solve operation.

Solution and check:

- $5 \cdot x - 2 \cdot (x-7) = 8 \cdot (x+2) - 17$

  $3 \cdot x + 14 = 8 \cdot x - 1$
- $(3 \cdot x + 14 = 8 \cdot x - 1) - 8 \cdot x$ $\quad 14 - 5 \cdot x = -1$
- $(14 - 5x = -1) - 14$ $\quad\quad\quad\quad\quad -5x = -15$

- $\dfrac{{}^-5 \cdot x = {}^-15}{-5}$ $\quad\quad\quad\quad\quad\quad x = 3$

- $5 \cdot x - 2(x-7) = 8 \cdot (x+2) - 17 \mid x = 3 \quad \text{true}$

Note: The notation shown above is slightly different from notation used to solve the problem on paper, but the steps are clearly shown. When the equation is entered, the CAS simplifies it on line 1. On line 2, the student subtracts $8x$ from each side. On line 3, the student subtracts 14. On line 4, the student divides by –5 to obtain the solution. On line 5, the equation is shown to be true when $x = 3$.

CHAPTER 13

Initiating Students into Algebra with Symbol-Manipulating Calculators

Tenoch Cedillo
Carolyn Kieran

IN A TECHNOLOGY implementation project in Mexico, students in seventh-, eighth-, and ninth-grade mathematics classes from fifteen schools were equipped with symbol-manipulating calculators. In this chapter, we describe the research that was conducted with the eighth-grade students in introducing them to algebra in a specially designed computer algebra systems (CAS) environment. After presenting a sample of the activities and their rationale, we discuss the conceptions that students developed for letters, expressions, and algebraic equivalence and their strategies for dealing with geometric patterns. We also briefly discuss what the teachers observed and how it affected them. Students' development of algebraic code as a "way of getting the calculator to do what they wanted it to do" proved to be a powerful and motivating approach to algebra learning and reflects the untapped potential of CAS in introductory algebra instruction.

CAS and ALGEBRAIC ACTIVITY FOR STUDENTS AGES 13–16

Models of Algebraic Activity

When thinking about the role that CAS) may play in introducing students to algebra, we might usefully consider the main components of school algebra. Kieran (1996) has developed a model in which the activities of school algebra are categorized and discussed according to three types: (1) generational, (2) transformational, and (3) global–metalevel mathematical. The first type of algebraic activity—*generational* activities—involves forming expressions and equations that are the objects of algebra. Typical examples include (*a*) equations containing an unknown and representing quantitative problem situations (Bell 1995), (*b*) expressions of generality arising from geometric patterns or numerical sequences (Mason 1996), and (*c*) expressions of the rules governing numerical relationships (Lee and Wheeler 1987). Much of the initial meaning-making of algebra is situated in these generational activities. The second type of algebraic activity—the *transformational*, or rule-based, activities—includes collecting like terms, factoring, expanding, substituting, solving equations, simplifying expressions, and so on. This type of activity involves the traditional symbolic manipulation that is commonly equated with algebra. The third type of algebraic activity—*global–metalevel mathematical* activities—uses algebra as a tool, but these activities are not exclusive to algebra. The global-metalevel mathematical activities include problem solving, modeling, noting structure, justifying, proving, and predicting; these activities do not necessarily require algebra. However, attempting to divorce these metalevel activities from algebra removes any context or need that students might have for using algebra. By engaging in global–metalevel mathematical activities, students are encouraged to develop an awareness of the role that algebra can play in mathematics.

CAS and Algebra Learning

Many people equate CAS use most directly with the transformational activity of algebra, but the use of the tool is not restricted to such activities. Waits and Demana (1998) state that CAS typically have the following basic capabilities: as computer symbolic algebra utilities for solving algebraic equations and manipulating algebraic expressions; as graphical software packages; as numerical solvers; and for exact arithmetic—rational, real, and complex. When CAS first emerged, they were viewed as helpful tools for mathematicians. Their main function was to do symbolic manipulation, sometimes within the context of a problem-solving activity. Dating as far back as 1982, Wilf (1982) foresaw the day when this tool would be available as a "pocket computer" in the hands of students. Since 1995 the availability of CAS extended beyond that of computer environments to include handheld calculators, such as the TI-89 and TI-92. However, as of 2002, these calculator-based symbol manipulators have yet to be widely used in high schools.

When CAS began to appear in college mathematics courses in the 1980s, most high school mathematics teachers saw little potential for their use in algebra classes because at this level of schooling, students were expected to perform algebraic transformations by hand. Gradually, however, the availability of this tool led some instructors to reconsider its use as a way to reduce the amount of time spent on symbol manipulation in high school algebra classes. However, educators were wary. For example, Fey (1989) voiced the concerns of many people when he asked whether "students can learn to plan and interpret manipulations of symbolic forms without being themselves proficient in the execution of those transformations" (pp. 206–7). Research has not yet answered that question decisively; however, some studies (e.g., Huntley et al. 2000) have shown that technology-supported algebra curricula that do not emphasize symbolic manipulation can lead to higher success rates on algebraic tasks embedded in applied problem contexts than curricula focusing principally on developing manipulative skills. Thus, students, with the aid of technology, can learn to plan and interpret symbolic forms without necessarily being themselves proficient. Zehavi and Mann (1999) have reported similar results from their study of year-8 students in Israel who used a CAS to solve equations after setting up the equation themselves. The researchers noted that students immediately checked to see whether the solution made sense; if not, they knew that the source of the trouble was the equation and not the algebraic manipulations.

We often hear of the "white box, black box" metaphor (Buchberger 1990) and the notion that white-box activity—that is, construction of meaning for the concepts and underlying algorithms of algebraic manipulation—is essential before using a CAS in a black-box activity for problem solving. For example, an Austrian study of 2000 thirteen- to seventeen-year-old students reported that "the greatest problem is that students who only use this powerful system ... as a large black box simply learn words and how to use these words but they do not learn problem solving by reasoning" (Heugl 1998, p. 19). A weakness of this metaphor is that it polarizes the discussion in terms of "before" and "after," with pupils using the black box simply as a tool for problem solving *after* they have learned to reason about the problem-solving situation. Implicit in the argument is the limited view of CAS merely as algebraic tools to be harnessed for use within the global metalevel activities of algebra described above. An alternative perspective is to ask about the potential of a CAS as a "gray box" environment—one that intertwines both the white and black boxes. Rather than talk in terms of white-box activity either before or after the black-box activity, we can focus on the kinds of algebraic learning that might take place during the period of use of the CAS and that are, in fact, occasioned by the CAS. Such a focus is reflected in this chapter, where the CAS is not treated exclusively as a problem-solving tool but is also considered as a mediator of algebra learning—a tool that helps create simultaneous meaning for the objects and transformations of algebra.

THE STUDY

Context of the Research

In Mexico, a four-year educational development project was launched in 1997 by which the Ministry of Education incorporated new technologies into basic education to provide access to modern scientific and advanced mathematical ideas (Moreno et al. 1999). The mathematical segment of the project included the following technologies: spreadsheets (Excel), Cabri-Geometry, SimCalc-Mathworlds, Stella (research version), and the TI-92 algebraic calculator. These various technological environments and tools were inserted into such curricular topics as arithmetic, prealgebra, algebra, geometry, variation, and modeling.

In fifteen secondary schools from around the country, thirteen- to sixteen-year-old students in seventh, eighth, and ninth grade participated in the project. Each school was equipped with some subset of the various computer technologies, along with the calculator technology. Thus, classes in all fifteen schools served as project sites involving the use of the TI-92s. The first author of this chapter was the local consultant in charge of all aspects of the calculator segment of the project, and the second author served as outside consultant. Teaching materials were developed, teachers were trained, ongoing support was provided throughout the course of the project, and research was conducted on various aspects of the implementation process.

The methodological approaches used in the educational development project as a whole incorporated both global and local levels of assessment. Although the global aspect of the research is not covered in this chapter, it focused on understanding the educational system as a complex model and was concerned with the evolution of educational processes including teachers, school principals, and parents as essential elements. The local level of assessment concentrated mainly on students' learning and longitudinal case studies related to the various computer and calculator technologies. The portion of the research that related to the TI-92 calculator segment of the project is the subject of the present chapter.

Methodological Aspects, in Brief

The major aims of the calculator segment of the study were to investigate the following:

- The role of CAS as a factor of change in secondary school teachers' conceptions and practices (seventh, eighth, and ninth grades)
- The ways in which the use of CAS might influence students' learning of, and motivation in, algebra

- The second aim, which is the main focus of this chapter, served as an indicator to refine the observations of teachers' ways of working and to attain more reliable evidence of the role played by CAS as a cognitive tool.

Because the Mexican curriculum includes five 50-minute sessions a week for a mathematics course, the teachers were asked to use calculators and the activities designed for the project in at least two out of those five sessions. The student data were drawn from the worksheets gathered from the eighth-grade classes of all the participating schools, from videotaped data and observations of the algebra classes collected by the researchers during their regular visits to the schools, and from informal student interviews throughout the three years of fieldwork. The project students were typical of those found throughout the country in schools where the average class size is about forty students. The worksheet materials, samples of which are presented subsequently, were developed specially for use with symbol-manipulating calculators (Cedillo 1997, 2001).

ACTIVITIES DEVELOPED FOR THE STUDY

Pedagogical Intent

Two guiding principles were used in the design of these activities: (1) take advantage of the CAS resources installed in the TI-92 calculator, resources that allow students to make a tight connection between arithmetic and algebra, and (2) serve as an introduction to school algebra that does not require students to master a priori any definitions and that helps them move from prealgebra to the use of algebraic code as a means to model and solve problems. These points are further discussed in the following paragraphs.

The CAS installed in the calculator permits working with algebraic expressions as "active objects" in the sense that the user can not only record the statement of a problem algebraically but also do something mathematically with such expressions and receive immediate feedback from the machine (see **fig. 13.1**). This feature of the calculator permits the student to enter an algebraic expression onto the working screen, including a list of trial values of the variable, and to see the results of the calculations that are carried out with the entered formula. In other words, given a number pattern, students can use the calculator to explore and verify their conjectures so as to define and edit an algebraic expression that represents the rule governing such a number pattern. In this way, we assume that they can start learning about the algebraic code by using it.

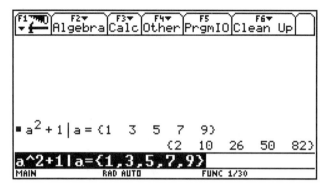

Fig. 13.1. The CAS provides immediate feedback on the mathematical manipulation of an algebraic expression.

The best example of learning through use is children's learning of natural language. The mother tongue is learned mainly through use, without previous knowledge of grammar or syntactical rules. To create an environment in which students meet algebra as a language-in-use, the project developers designed a pedagogical approach by recasting Bruner's (1982, 1983) research on language acquisition. The notion of algebra as a language is pervasive in the mathematics education literature. Thus, algebra can be viewed as a linguistic code for communicating with the computer or calculator.

According to Bruner (1982, 1983), language often proliferates, or at least is more forthcoming, when children are in "playful" situations. We therefore decided to create a gamelike activity based on recognizing and producing number patterns using the calculator code. On the one hand, this type of activity allows us to put students in the position of meeting algebra without previously knowing rules and definitions. On the other hand, it also provides the mathematical structure of linear functions for outlining a route from prealgebra to modeling and solving algebra problems. In addition, the use of the calculator was guided by the principle of expressing generality by means of the calculator code. On the basis of this principle, we created a mathematical environment in which pupils meet algebraic code as a language that helps them explore the behavior of number patterns, produce algebraic expressions that encapsulate and govern them, and use the algebraic expressions they built as tools to answer questions regarding those number patterns. The tight link between general numerical relationships and the production of algebraic expressions mediated by the calculator is meant to constitute a shared context between the students' previous arithmetical experience and the new algebraic code.

Letters and algebraic expressions were introduced on the worksheets as a means of editing programs with a calculator. Initially, as far as the students were concerned, a letter was not an unknown or a variable but a label denoting the name of a memory that the calculator used to store the information they had entered. Similarly, an algebraic expression was a calculator program that consisted of a chain of operations that included the name of a memory represented by a letter. In the activities of the study, the students were asked to describe the rule for a linear-based number pattern and to use the symbolic features of the calculator to create a "program" (expression) that reproduced the pattern. For example, the rule that generates the pattern in the following table can be expressed in the calculator as A∗2−1.

| Input number | 1 | 4 | 6 | 9 |
| Output number | 1 | 7 | 11 | 17 |

The students described verbally the rule that generates this table as "multiply by 2 and subtract 1," or "add the input number to itself and subtract 1." After they programmed the calculator with a linear function that they thought represented this rule, they assigned numerical values to the variable to validate their programs. This process would be represented in the CAS by an entry like A∗2−1|{1, 4, 6, 9} with resulting output {1, 7, 11, 17}. We wish to emphasize that no instruction was given as to nomenclature or algebraic concepts. As far as the students were concerned, the algebraic expressions were merely programs that allowed the calculator to "understand" what they wanted it to do.

The experimental activities were organized into six blocks containing ten to fifteen worksheets each. Throughout the blocks, the students confronted different mathematical activities whose content gradually changed. Each worksheet included a new element, be it numerical, involving a sign or decimal point, or structural, such as a one-step rule (e.g., 3×D) or a two-step rule (e.g., 3×D+1). These activities required students to use the calculator language in the following ways:

- Block 1: describe number patterns
- Block 2: produce number patterns
- Block 3: produce equivalent algebraic expressions
- Block 4: describe part-whole relationships algebraically
- Block 5: explore inverse linear functions
- Block 6: confront problem situations that can be solved by creating an algebraic model

Examples of the Activities

In this section, we use four representative activities to illustrate how the study of algebra was introduced in eighth-grade classes participating in the project. The order in which the activities appear corresponds to the sequence used in the classroom. The activities—viewed from both teachers' and students' perspectives—will be discussed later in this chapter.

Example 1: Initiation into the use of algebraic language (block 1)

I made the following table using a program:

Input number	1.1	2.6	3	4.3	5
Output number	3.2	6.2	7	9.6	11

- What result will the calculator give me if I type the number 50 into my program? If I type the number 80? If I type the number 274?
- What operations did you carry out to obtain these results?
- Can you program your calculator to do the same thing as mine? Write your program below.
- Use the program you made to find the numbers missing from the table.

Input number	17	35.02	89.73	107.06	299.10	307.09		
Output number							511	613.03

- What operations did you carry out to obtain the values associated with 511 and 613.03?

Example 2: Introduction to algebraic equivalence (block 3)

I made a program that does the following:

Input number	2	4	8	10	14
Output number	3	6	12	15	21

- If the input number is 5, what result will the calculator give? And if the input number is 6? If the input number is 15? What operations did you carry out to obtain these results?
- Can you program your calculator to do the same? Once you have written the program, check it with your calculator; if it works, write it below.
- A pupil says that the program $b + b \div 2$ gives the same results. Do you agree? Write down two examples to justify your response.

- Can you make another program, like the one she made, that will also produce the same results as those shown in the table? Check it with your calculator; if it works, write it down next.

Example 3: Algebraic representation of the part-whole relationship (block 4)

In my grandfather's hardware store are rolls of wire sold by the kilo. All the rolls weigh the same. To help him, I made a program that, if I input the quantity sold, gives a result that indicates how much wire is left.

Wire sold (kg)	1.7	2.4	3.1	4.06	5.2
Wire remaining (kg)	8.3	7.6	6.9	5.94	4.8

According to the information provided by the table, how many kilos of wire are in a roll?

Can you make a program that does the same as mine? Check it with your calculator, and write it down below.

Use your program to complete the following table.

Wire sold (kg)	2.83	3.03	3.5	4.8				
Wire remaining (kg)					5.01	6.2	7.04	7.32

How can you check the values you found for 5.01, 6.2, 7.04, and 7.32? Explain your method in a way that your classmates would understand.

Example 4: Problems involving linear functions (block 6)

Look at the following pictures, and draw the next two pictures.

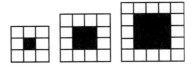

- How many squares are needed to make the frame of the black square in picture number 27?
- How many squares are needed to make the frame of the black square in picture number 100?
- Explain the reasoning behind your answers to the two previous questions.
- Can you program your calculator to complete the following table? Write down the program you made.

Place of the picture in the sequence	48	75			
Number of square needed for the frame			704	772	840

Guidelines Given to the Teachers for Implementing the Activities and the Technology

The recommendations offered to the teachers for implementing the activities and the calculators are summarized in the following paragraphs. The ways in which they put these recommendations into practice are reported in a subsequent section in this chapter.

The classroom setting consisted of the following elements:

- A teacher who fulfils the role of proficient user of the calculator's language, fine-tuning the way in which it is used so as to fit with students' present level of knowledge

- A group of approximately forty students twelve to thirteen years of age, each possessing a calculator with a new sign system that can express arithmetic procedures algebraically

- A set of previously designed worksheets that structure the students' uses of the calculator's language to make the calculator "do their bidding" by using "words" from the calculator's programming code that the calculator "understands"

Work begins by showing the students how to write an algebraic expression program for the calculator and what the program does when it runs. The activity consists of a gamelike task in which the students "guess" another student's program. Pupils are given clues, such as a numeric pattern shown in a table, that guide them to produce conjectures. They are then challenged to program the calculator to produce this table. Students win the game if they express the rule as a program for the calculator in such a way that they reproduce the given numerical pattern by running the program, as shown in example 1, block 1 in the previous sections. The teacher's objective is to achieve as many winners as possible without trivializing the task. To do so, the teacher asks students to raise their hands when they think their work is done, then checks their work individually. Every two weeks, the teacher organizes sessions to collectively discuss the group's achievements and difficulties.

At the end of the class, the students are asked to hand in worksheets they have completed so that the worksheets can be marked and returned to them before the next lesson. The marking consists of brief written comments on the student's answers. If a worksheet is completed incorrectly, no corrected answer is given to the student; instead, the teacher points out that the work is wrong and poses a new question that can assist the student in recognizing the mistake. If the worksheet is completed correctly, the teacher adds a new question to motivate the student to find another way of solving the problem, encouraging the student to learn more.

RESULTS AND DISCUSSION OF STUDENT LEARNING

This section focuses on key student achievements during the fieldwork of the 1997–2001 development project in Mexico. The report is based on students' responses to activities similar to the activities illustrated by examples 1, 2, and 3 herein. The work of several hundred students from three different regions of the country, amplified by researchers' classroom observations and questions, served as the principal source of data for this report.

Notions Developed by Students for Letters and Algebraic Expressions

The activities in block 1 were designed to introduce students to the algebraic code by having them use it without any previous knowledge of rules and definitions. According to the aims of this study, the following specific features were observed:

- The mathematical meanings that students assigned to letters and the algebraic expressions they used to program the calculator
- Students' strategies in using the calculator code to negotiate problem solutions

During classroom observations, the two following questions were consistently put to students:

1. When you create a program in your calculator, what does the letter you use mean to you?

2. What does a program like the ones you have created in your calculator mean to you?

The following vignette in response to question 1 encapsulates the type of answer given by the majority of the students.

> The letter I use in a program serves to make the calculator recognize any number I input…. As many numbers as you want…. May I use the calculator? (types the program a∗2+3)…. If I input 5, the program gives 13, that is … 5 two times plus 3…. See ,… if I input 9, it does the same … doubles 9 and adds on 3…. And so on, it keeps doing the same operations I order the calculator to do, no matter which number you input.

And to question 2, the following type of answer was given by most students.

> I use a program to do something, ... for example, to complete a table.... Sometimes I make a program to solve a problem.... A program helps me to find out how to solve a problem.

The answers shown suggest that students were developing the notion of a letter as a symbol "representing any number" and the notion of an algebraic expression as a "computing device." The next episode provides additional evidence for this assumption. During a visit to a school, the following situation was posed to the students: "Think of an integer between 0 and 10, take away your number from 10, and keep the result. Next add 10 to your initial number, and keep the result. Finally add the two results you have. May I predict the final result that you got? It is 20. Can you explain why I was able to find the final number you got?"

After a few moments of reflection, the students produced several explanations. The first attempts were rather cumbersome until a pupil said, "You can anticipate the result because you had made a calculator program.... No matter which number we thought of, you made a program that gives always the same." Another pupil chimed in, "I've got that program," at the same time that many other students were putting up their hands to answer the question. Then students were invited to individually show their programs to the visitor to give the rest of the students a chance to think about the problem. Not only were these students able to produce a program with the calculator to represent the relationships involved in the puzzle, they also made discoveries while exploring with different values, as illustrated below by the words of one of the students:

> It will always give 20, no matter which number you input;...the number can be a fraction or a negative number,...because you have $b + 10 + 10 - b$,...see,...b is the number you thought of, but the b's add nothing....It is like just having 20, then you add and take away b from 20, that is why it will always give 20,...no matter which number you want b to be.

Worth emphasizing is the point that no rules or definitions were given to the students during the teaching phase. The students' reactions provide evidence for the claim that the meanings they attached to these algebraic entities were the result of facing algebraic code through using it. We should also remember that, for the students taking part in this study, an algebraic expression was a calculator program; when they referred to a "program," the word *program* meant an algebraic expression. The students' responses suggest that an essential aspect of the development of these notions was that the calculator allowed them to use a programming code not just as a means of editing but also as a language for negotiating problem solutions: "You can anticipate the result because you had made a calculator program.... No matter which number we thought of, you made a program that gives always the same."

Another influential factor was that the activities were based on both the description and production of numerical patterns, thus helping students establish a connection between the new formal code they were using and the arithmetical experience they had acquired in previous courses. The strong relationship between the experimental tasks and the calculation tool allowed them to use the numerical referent as a way of validating their responses. This outcome resonates with Bruner's research (1982, 1983). According to Bruner, the tight relationship between context and the child's speech makes it possible for the child to start making sense of language. In this study, such context is provided by the link between the immediate feedback given by the calculator and the students' arithmetical background.

The data suggest that because the calculator code is placed within the calculator's computing environment, it evoked in students the idea that algebraic expressions are "expressions for calculating." The numerical strategy of "trial and refining," which the students used to validate or refute the algebraic expressions they produced, gave evidence of this perception. The availability of calculators permits activities involving recognition of numerical patterns to go beyond encoding what is represented, as happens when working with paper and pencil. In the calculator-based environment, algebraic symbolization is the result of an interaction between what is known (arithmetic) and the process of attaining a goal (making the calculator reproduce a table of given values). For example, the program 3xB-1 generates the numerical pattern 2, 8, 17, 23 when B = 1, 3, 6, 8. If this pattern was the one that students wanted to generate, they knew that the program they had written was correct; if not, they could reanalyze the pattern and try again.

The data from this research suggest that this type of activity helps students step from the particular to the general—from analyzing the behavior of a specific ordered pair $a \rightarrow b$ to verifying the validity of the rule that they found so they could apply it to any pair $x \rightarrow y$ that could be in the table. The ways the students worked indicate that this experience was crucial to their development of the instrumental notion of a letter as "serving to represent any number" and of an algebraic expression as being a "thing that is used to do something ... to complete a table or solve a problem."

Students' Notions of Algebraic Equivalence

The students were asked two pivotal questions to delve into the extent to which the use of calculator code may have helped them develop some initial notions about algebraic equivalence. They were asked these two questions:

1. "A pupil from another school says that the programs $(a+7) \div 2$ and $(z+7) \div 2$ produce different results. What do you think about it?"

2. "A pupil from another school says that $(a + b)^2 = a^2 + b^2$. What do you think about it?"

The students' responses to these questions suggest that they were able to develop well-formed notions concerning algebraic equivalence on the basis of their explorations with the numerical value of the algebraic expressions. This strategy marks a clear relationship between these notions of equivalence and the use of the calculator code to describe number patterns. About four-fifths of the students gave answers that can be summarized by the following excerpts.

To question 1 above, one student responded as follows:

> These two programs give the same results because when you input a number, what matters is the operations you want to perform with such number.... The calculator understands the same whether the program is $(a+7) \div 2$ or $(z+7) \div 2$.... I have done it before, it can be a or z or any other letter; if you maintain the same operations, the calculator understands the same.

In answering question 2 above, almost all the students were able to give correct answers by using examples, a phenomenon that seems to be related to the instrumental meaning attached to algebraic expressions mentioned previously. Some of these students went further by finding that "$(a + b)^2 = a^2 + b^2$ if $a = 0$, $b = 0$, or if both are zero." These students' responses suggest that they had developed the notion that different letters in an algebraic expression can represent different values but can also have the same value. The students' notions of algebraic equivalence can be synthesized by the following response of a student to question 2 above: "Two programs are equivalent if they produce the same values."

Approximately 90 percent of the students gave correct answers to questions 1 and 2 above. This outcome contrasts with the results obtained by Küchemann (1981) in the large-scale CSMS study where students were working with paper and pencil. Although the students in the present research associated literal terms in algebraic expressions with numerical values, Küchemann reported that a significant number of the CSMS students had attached meanings to letters that were linked with their previous experience in geometry and measurement or with their use of everyday written language, a conclusion that seemed to work against them in achieving correct notions for algebraic equivalence.

The students who participated in this study were free to choose their favorite letter from the calculator keyboard to create a program, possibly instilling the notion that the numerical value of an algebraic expression does not depend on the letter they use. The question about the equivalence between $(a + b)^2$ and $a^2 + b^2$ presented the students with a more complex situation. The activities used in this study did not provide any direct experience working with two variables and quadratic expressions; yet the students attempted to respond to this question with the same numerical strategies they had used with linear expressions. Lee and Wheeler (1989) asked the same question aimed at investigating the extent to which students, in their last year of high school, were aware of a connection between arithmetic and algebra. They reported that many of their students were not able to give correct answers using either algebra or arithmetic and that only a few of them looked for numeric examples to handle the question. In this respect, the students' responses in the present research underlined the possibilities that arise when working with symbolic calculators in this didactic approach.

Students' Strategies in Confronting Geometrical Patterns

Geometrical patterns, such as the one presented in the section of sample activities above, helped students generate a range of different strategies. The most frequently used strategy was to build a table like the one shown below. Approximately half the students used this approach. The table guided them in finding the number pattern and in eventually producing a calculator program to solve the problem.

Place of the picture in the sequence	1	2	3	4	5
Number of squares needed for the frame	8	12	16	20	24

The fact that many students used this approach is not surprising, because the way in which the activity was presented may have induced the idea of making a table with the numerical values they observed in the geometrical sequence.

The second most frequently used strategy, used by approximately a fourth of the students, was to treat the sequence as a "shifted list" of multiples of a given number and to combine this treatment with a trial-and-refining approach to negotiate a solution for the problem of finding "how many white squares are needed to make the frame of the black box in picture number 100?" For example, by inspection, they saw the geometrical pattern shown in the figure below as "multiples of 4." The following extract from a conversation with a student illustrates this observation:

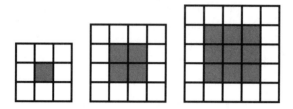

> The first figure in the sequence has 8 white squares; the second, 12; the third, 16.... It looks like the multiples of 4.... Thus, the program should be something like k∗4,... but if k = 1, it gives 4, and I need 8; if k = 2, it gives 8, but I wanted 12.... Thus, the program must be k∗4+4.

Approximately a tenth of the students related the sequence to the geometric concept of area. Such a strategy is illustrated in the following statement by a student who created the program (a+2)² − (a×a), where

> *a* is the length of the black box, and *a* + 2 is the length of the big square.... $(a + 2)^2$ computes the area of the big square, and $(a*a)$ computes the area of the black box; the difference is the number of white squares surrounding the black box.... To calculate how many white squares are needed to make the frame of the black box in picture number 100, I ran the program I made for *a* = 100.

The various strategies discussed above suggest that the support provided by the calculator induced students to view the algebraic code as a tool with which to reason. This notion was developed by students' exploring and refining their conjectures instead of simply using algebra to express ready-made, polished ideas. Less than a tenth of the students used a strategy where symbolizing was not preceded by trial-and-refining explorations. Thus the majority experienced an explicit intertwining of the numerical and the algebraic, of the particular and the general. But, perhaps more important, this finding highlights the action orientation underlying a "programming" approach to introductory algebra, whereby algebraic expressions are not perceived as static representations of various phenomena but as tools for calculating. This finding thus resonates with earlier research (Clement 1982) that, even though it was not conducted with CAS, showed that teaching students to view equations in terms of numbers on which specific operations are performed was more productive than encouraging them to see equations simply as structural representations. The advantage of the CAS environment, as designed for this study, is that it seemed to induce naturally such an action perspective on algebraic expressions.

WHAT THE TEACHERS SAW AND HOW IT AFFECTED THEM

Our students' reactions and our classroom observations were evidence that the introduction of symbol-manipulating calculators in the classroom was a crucial factor in promoting not only students' learning of algebra but also their positive attitude toward mathematics. Approximately 75 percent of the students declared that they had not liked mathematics before using calculators, "because it was boring and they could not understand the teacher's explanations." Once students began using calculators to confront the new ideas, many students having previous histories of failure found that "with time, they could be as good at mathematics as those students who had always been acknowledged as the best in the school." Teachers' comments confirmed the students' reactions, in both individual interviews and meetings. Teachers consistently and persistently spoke of "their surprise at how students with a weak mathematical background were progressing in their mathematical reasoning." This statement does not suggest that the stronger students were not also advancing in their learning of algebra in the CAS environment but, rather, that the teachers were particularly interested in observing the outcomes of using the technological approach with the students who had been more difficult to reach.

Students' mathematical progress produced a very important side effect, one that involved the teachers themselves. Among the teachers taking part in the project were those whose mathematical background was quite strong, yet who seemed the most difficult to convince about the benefits of using technology. Particularly those with a long experience in teaching proved to be the most reluctant, seeming to perceive that their experience and knowledge of the discipline precluded the need to learn new pedagogical approaches. Classroom observations and videotapes during the first year of the project revealed that they were using the calculators in a rather halfhearted manner.

The improvement in students' attainment seemed overtly to be the most influential factor in changing teachers' initial view of technology. Once they witnessed that their students were using strategies not taught in class yet were progressing well, they started to change their attitudes and rapidly became avid appreciators of the new possibilities provided by technology. Change was not always quick, however, taking a year for one teacher and almost two years for another.

The teachers who changed their attitudes also changed their teaching styles. They abandoned lecturing at the front of the classroom. They started to produce new materials on their own to supplement the activities they had been provided by the project. They began to respect students' self-pace of learning and to be alert to helping students on an individual basis; their help encompassed both the more advanced students so that they might reach a higher level of attainment and the weaker ones so that they might be assisted in their areas of difficulty.

Even though teacher change was not the main focus of this chapter, we would have been remiss in discussing the role played by the symbol-manipulating calculator in introducing students to algebra had we not also brought to light the unexpected benefits with respect to the teachers. Seeing their students succeed with this technology was essential in their metamorphosis.

CONCLUSION

Many teachers—not those teachers involved in the project—seem to resist the idea of using CAS to introduce their students to algebra. They believe that students should learn symbol manipulation by hand before being permitted to use a CAS tool to carry out symbolic work. The evidence presented in this chapter shows that hundreds of students who were introduced to algebra in a specially designed CAS environment learned some elements of symbol manipulation simultaneously with describing and producing formal algebraic code. Although their initiation did not encompass a full range of transformational activity, it provided a kind of existence proof as to the rich meanings that can be fostered at the same time students are learning a language with which they might communicate mathematically with a calculator or computer. The manipulative aspect of algebra thus becomes reconceptualized as "getting the calculator or computer to do what you want it to do."

In the introductory section of this chapter, we described the activities of algebra as generational, transformational, and global. In the research presented in this chapter, all three components were intertwined. Students' description of numerical situations, production of algebraic code, and experience with equivalent forms were all tightly related. Furthermore, the students in this study did not first engage in white-box, conceptually oriented activity prior to using the CAS as a black-box tool for problem solving. Instead, they developed their conceptions of algebraic objects and some simple transformations of these objects—such as simplifying expressions, inversing functions, and producing equivalent expressions—all at the same time they were communicating with their symbol-manipulating calculator. We have chosen to call this a "gray-box approach" to CAS use. From both the teachers' and the students' perspectives, it proved to be an effective way of introducing students to algebra.

Our final comment is a return to the finding related to student achievement and motivation. When teachers saw their students' improvement and increased interest in mathematics, they quickly became convinced of the pedagogical advantages of using CAS for all students, including those they considered weak in mathematics. It seems reasonable that students who are interested in mathematics stand to learn more than those who are not. However, from our perspective, the CAS was more than a motivational artifact. It permitted students to be introduced to algebra in such a way that the symbol-manipulating aspects were tightly tied to the students' prior numerical experiences in arithmetic, thus making algebra a more meaningful activity for students.

ACKNOWLEDGMENTS

We want to thank Teresa Rojano of CINVESTAV and Elisa Bonilla of the Mexican Ministry of Education for their co-direction of this project, as well as Sonia Ursini, who coordinated the EMAT subproject; Marcela Santillan, from ILCE, who facilitated the numerous technical aspects associated with the project; and all the other members of the research teams in CINVESTAV, SEP, UPN, and ILCE. Funding for this research was also provided by grants from The National Council of Science and Technology, Mexico, reference G26338-S and 30523-S.

REFERENCES

Bell, Alan. "Purpose in School Algebra." *Journal of Mathematical Behavior* 14 (1) (1995) (Special Issue: New Perspectives on School Algebra: Papers and Discussions of the ICME-7 Algebra Working Group, edited by Carolyn Kieran): 41–73.

Bruner, Jerome S. "The Formats of Language Acquisition." *American Journal of Semiotics* 1 (1982): 1–16.

———. *Child's Talk: Learning to Use Language*. New York: W. W. Norton & Co., 1983.

Buchberger, Bruno. "Should Students Learn Integration Rules?" *Sigsam Bulletin* 24, no. 1 (1990): 10–17.

Cedillo, Tenoch E. *Calculadoras: Introducción al Algebra* (Cuadernos Didácticos, Volumen 4). México: Grupo Editorial Iberoamérica, 1997.

———. "Towards an Algebra Acquisition Support System: A Study Based on Using Graphic Calculators in the Classroom." *Mathematical Thinking and Learning* 3, no. 4 (2001): 221–59.

Clement, John. "Algebra Word Problem Solutions: Thought Processes Underlying a Common Misconception." *Journal for Research in Mathematics Education* 13 (January 1982): 16–30.

Fey, James T. "School Algebra for the Year 2000." In *Research Issues in the Learning and Teaching of Algebra* (vol. 4, Research Agenda for Mathematics Education), edited by Sigrid Wagner and Carolyn Kieran, pp. 199–213. Hillsdale, N.J.: Lawrence Erlbaum Associates, and Reston, Va.: National Council of Teachers of Mathematics, 1989.

Heugl, Helmut. "The Influence of Computer Algebra Systems in the Function Concept." Paper presented at the International Conference on Technology in Collegiate Mathematics, New Orleans, November 1998.

Huntley, Mary Ann, Chris L. Rasmussen, Roberto S. Villarubi, Jaruwan Sangtong, and James T. Fey. "Effects of *Standards*-Based Mathematics Education: A Study of the Core-Plus Mathematics Project Algebra and Functions Strand." *Journal for Research in Mathematics Education* 31 (May 2000): 328–61.

Kieran, Carolyn. "The Changing Face of School Algebra." In *Eighth International Congress on Mathematical Education: Selected Lectures*, edited by Claudi Alsina, José Alvarez, Bernard Hodgson, Colette Laborde, and Antonio Pérez, pp. 271–90. Seville, Spain: S.A.E.M. Thales, 1996.

Küchemann, Dietmar. "Algebra." In *Children's Understanding of Mathematics 11–16*, edited by Kathleen Hart, pp. 102–19. London: Murray, 1981.

Lee, Lesley, and David Wheeler. *Algebraic Thinking in High School Students: Their Conceptions of Generalisation and Justification* (research report). Montreal: Concordia University, Mathematics Department, 1987.

———. "The Arithmetic Connection." *Educational Studies in Mathematics* 20, no. 1 (1989): 41–54.

Mason, John. "Expressing Generality and Roots of Algebra." In *Approaches to Algebra: Perspectives for Research and Teaching*, edited by Nadine Bednarz, Carolyn Kieran, and Lesley Lee, pp. 65–86: Dordrecht, The Netherlands: Kluwer Academic Publishers, 1996.

Moreno, Luis, Teresa Rojano, Elisa Bonilla, and Elvia Perrusquía. "The Incorporation of New Technologies to School Culture: The Teaching of Mathematics in Secondary School." In *Proceedings of the Twenty-first Annual Meeting of PME-NA*, Vol. 2, edited by Fernando Hitt and Manuel Santos, pp. 827–32. Columbus, OH: ERIC, 1999.

Waits, Bert K., and Franklin Demana. "The Role of Hand-held Computer Symbolic Algebra in Mathematics Education in the Twenty-first Century: A Call for Action!" Paper Presented at NCTM Standards 2000 Technology Conference, Washington, D.C, June 1998.

Wilf, Herbert S. "The Disk with the College Education." *American Mathematical Monthly* 89, no. 1 (1982): 4–8.

Zehavi, Nurit, and Giora Mann. "The Expressive Power Provided by a Solving Tool: How Long Did Diophantus Live?" *International Journal of Computer Algebra in Mathematics Education* 6, no. 4 (1999): 249–66.

Chapter 14

Algebra on Screen, on Paper, and in the Mind

Paul Drijvers

IN THIS chapter, I investigate the relationship among computer algebra use, traditional paper-and-pencil work, and algebraic thinking by interpreting students' behavior from a theoretical perspective. The main elements of this theoretical framework are the process-object issue and the notion of instrumentation. The data stem from an educational experiment in ninth-grade classes in which the students used a symbolic calculator to solve algebraic problems. The computer algebra was used for solving (parametric) equations and for substituting expressions. I make the observation in this chapter that the work in the computer algebra environment is closely related to paper-and-pencil work and to students' mental conceptions. The theoretical standpoints offer ways of making this connection more explicit and better understanding students' difficulties.

Introduction

Computer algebra systems, both handheld and desktop versions, are "finding their way" into mathematics classrooms on a regular basis. In the early 1990s, technology became more widespread and software became more user-friendly. The interest in the potential of computer algebra for mathematics education grew, and optimism dominated the debate. This technological device began to open new horizons that had previously been inaccessible and provided opportunities for exploring mathematical situations. In this way, the tool facilitated students' investigations and discoveries.

The optimism about CAS is the classroom has taken on additional nuances. The research survey of Lagrange and his colleagues (2001) indicates that difficulties arising while using computer algebra for learning mathematics have gained considerable attention. For example, Drijvers (2000a) addresses general obstacles that students encounter while working in a CAS environment. Tall and Thomas (Tall and Thomas 1991, Tall 1992) investigated prerequisite knowledge that students need for using the computer to enhance algebraic thinking. Heck (2001) specifically lists differences between the algebraic representations found in the computer algebra environment and those encountered in traditional mathematics.

The integration of computer algebra in mathematics education seems more complicated than we would like. As we consider the learning of mathematics primarily as the construction of mental mathematical objects, we naturally assume that a fruitful role of computer algebra in the learning process can be achieved only if the work in the computer algebra environment is congruent with students' mental images. For most students, paper-and-pencil work remains the primary medium for doing mathematics. Therefore, we think that students' mental mathematical objects and the paper-and-pencil work are often closely related. We need to ask ourselves the question, "How does computer algebra fit in?" In this chapter, my goal is to investigate the complex relationship among computer algebra use, traditional paper-and-pencil work, and algebraic thinking and to consider the impact of adding computer algebra to the traditional paper-and-pencil environment.

The following section of this chapter discusses the essential elements of the theoretical framework, the process-object duality, and the notion of instrumentation. Next, the educational setting is sketched and exemplary episodes of students' behavior in solving parametric equations are described. I use the theoretical perspective to reflect on these episodes. The specific characteristics of the solve command in the computer algebra environment are addressed. In a similar way, examples of students' reactions to exercises concerning algebraic substitution are described later in the chapter. The chapter ends with a discussion and a conclusion. The data in this chapter stem from an ongoing research project called "Learning Algebra in a Computer Algebra Environment" funded by The Netherlands Organization for Scientific Research (NWO) under number 575-36-03E (Drijvers 2001).

Theoretical Framework

Reflecting on the "triad" of CAS use, paper-and-pencil work, and algebraic thinking, we next look at two theoretical notions of particular interest: (1) the process-object duality and (2) the theory of instrumentation.

I first address the process-object duality and then discuss the concept of instrumentation, linking both theoretical perspectives with the classroom episodes that follow. By making the theoretical frameworks more concrete, we can investigate ways of using them to better understand observed student behavior.

Other perspectives that are relevant to the topic of this chapter are not addressed in this chapter. For example, no space is devoted to the classroom environment, social norms, mathematical norms, or the classroom culture. In this chapter, I do not investigate the changes over time due to the introduction of computer algebra, nor do I discuss the role of the teacher: How do teachers orchestrate the introduction of computer algebra into the mathematics lessons? How does the "didactical contract" change under the influence of new technological tools? At the risk of having too narrow a view, I confine myself to the two perspectives described below.

Process and Object

During the last decade, educational research has paid much attention to the dual character of mathematical conceptions. A mathematical conception is sometimes defined as a process that can be carried out, whereas in other situations it is more appropriately considered as an object that can be operated on by processes at a higher order. Gray and Tall (1994) elaborated on this idea in the notion of the *procept*. *Procep*t is a conjunction of *process* and *concept*; a procept is described as "a combined mental object consisting of a process, a concept produced by that process, and a symbol which may be used to denote either or both." Gray and Tall (1994) stress the flexibility that mathematics students need to deal with the ambiguity of mathematical notations. In "3 + 5," for example, the + may be an invitation to perform the process of addition or counting, but "3 + 5" can also be the object that is the result of the addition. It can be an object that must be squared to calculate $(3 + 5)^2$. In algebra, one usually cannot calculate $a + b$, and therefore the + is only a symbol that defines the object as "the sum of a and b." The mathematical expertise to decide whether the ambiguous notation in a specific situation denotes the process or the object is very important in algebra.

When students learn a new concept, the process aspect tends to be dominating. Students are inclined to calculate, to perform a procedure, to learn the algorithm—in other words, to *do*. In a later phase, the object aspect must be added and the "two sides of the coin" are integrated into one concept. Sfard (1991) uses the word *reification* for the "objectivation" of processes, a concept that is similar to the idea of encapsulation described by Dubinsky (1991). Dubinsky states that encapsulation of processes into objects is an important step in reflective abstraction.

Because the process-object duality is very manifest in algebraic concepts and operations and because computer algebra use influences the way of dealing with this issue, it seems to be a valuable theoretical perspective.

Tool and instrument

As a second theoretical perspective to analyze the student "episodes" that follow, I address the tool-instrument dimension, or the theory of instrumentation. The idea of instrumentation emanates from cognitive ergonomics (Rabardel 1995) and is related to using technological tools in learning. French mathematics educationalists (Artigue 1997, Lagrange 1999abc, Guin and Trouche 1999, Trouche 2000) have applied this notion to the learning of mathematics using information technology (IT), as will be done in this chapter, too.

An essential idea in the theory of instrumentation is that a tool is not automatically a useful instrument. For example, a hammer may be a meaningless thing unless you have used it before or have seen someone hammering. Only when a person wants a hammer and gains experience using it does the hammer become a valuable and useful instrument. The user develops skills to use it in a handy way and learns when the hammer is useful.

What holds true for the hammer can also be applied to other tools, such as calculators and computer algebra systems. Generally speaking, the learning process in which a bare tool or "artifact" gradually develops into a useful instrument is called *instrumental genesis*. Often, this process requires time and effort. The user must develop skills for recognizing the tasks for which the instrument is appropriate and must perform these tasks with the tool. The "how" of this process is condensed in the form of instrumentation schemes. The development of these schemes is an important aspect of instrumentation. An instrumentation scheme consists of two components: a technical part and a mental part. The technical part concerns the sequence of actions that a person performs on a machine to obtain a certain goal. In the use of mathematical IT tools, the mental part consists of the mathematical objects involved and a mental image of both the problem-solving process and the machine actions. Such mental mathematical conceptions are part of the instrumentation scheme and can evolve further during the development of the scheme. Technical skills and algorithms and conceptual insights are inextricably bound up with each other in the instrumentation scheme.

Exactly this combination of technical and conceptual aspects within the instrumentation scheme makes the theory of instrumentation promising while considering the relationship among computer algebra use, paper-and-pencil work, and mental conceptions.

Educational Setting

The experiments from which the observations in this chapter are drawn took place during spring of 2000 in two classes of about twenty-five students each and during spring of 2001 in two other classes of similar size. The students were in ninth grade, ages 14 to 15, and were preparing for the highest, precollege level of upper secondary education. Streaming, which is similar to "tracking" on the basis of future career interest, had not yet taken place, so we worked with future students of the exact stream and the language or society stream together. After the ninth grade, these students will be divided into different "profiles."

The experiments ran for five weeks with four weekly lessons of forty-five minutes each and dealt with ways in which computer algebra could support the development of insight in the conception of parameter. In this chapter, I concentrate on two types of algebraic operations: solving systems of equations that may contain parameters and substituting expressions into equations.

For a better understanding of the episodes or "scenarios" described below, I need to explain the approach of algebra in the Netherlands. The introduction of algebra during the first years of secondary education for students aged 12 to 15 takes place carefully. Much attention is paid to the exploration of realistic situations, to the process of *mathematization*, and to the development of informal problem-solving strategies. Variables, for example, are often words that have a direct relationship with the context from which they come. The translation of the problem situation into mathematics is an important issue, whereas formalization and abstraction are delayed. As a consequence, such algebraic skills as formal manipulations are developed relatively late and to a limited extent. Hence, at the time of the experiments described in this chapter, the students' knowledge of algebraic techniques was limited. For example, the general solution of a quadratic equation had not been taught yet.

The students participating in the experiment had access to the TI-89 symbolic calculator as their computer algebra tool. The rationale for this choice lies in the portability of the machine, its flexibility, and the soundness of the algebra module. The students used the TI-89 both in school and at home. Because these students had no previous experience with such technology as graphing calculators, using this type of handheld technology was very new to them.

The experiments began with a pretest. The students then worked through a booklet called *Introductie TI-89* (Introduction to the TI-89) (Drijvers 2000b) that was developed for this occasion. The booklet was designed to acquaint the students with the machine and prepare them for the relevant machine techniques. The "main dish" consisted of the booklet *Veranderlijke Algebra*

(Variable Algebra) (Drijvers 2000c), which targeted development of the notion of parameter. The assignments in this chapter come from these booklets. The experiments concluded with a posttest and with interviews with a select group of the students participating in the experiments.

The TI-89 Solve Command versus the Notion of Solving

To illustrate the relationship among mental mathematical conception, paper-and-pencil tradition, and computer algebra use, we first consider the solution of equations that may contain parameters. The episodes of student behavior in the next two sections illustrate the differences between the solve command as it is available in the TI-89 calculator and the way students perceive the term "solve" from the traditional paper-and-pencil setting. The computer algebra environment seems to emphasize elements in the solving process that often are implicit in solving "by hand" or "by head."

Solving a Parametric Equation

The TI-89 solve command is introduced to the students in the introductory part of the teaching material. The first exercise concerning solve is simply to enter the command:

$$\text{Enter: solve}(3x+5 = 7, x)$$

The students are told that after the command "solve" is entered, they should give the equation, then a comma, and finally the letter with respect to which they want to solve the equation. The latter aspect is new to them because so far the unknown has always been implicit. In most situations, they have not had to choose the unknown because the equation has contained only one variable.

The next assignment is formulated in mathematical language instead of machine language:

$$\text{Solve: } x^2 = 3$$

Note that the difference between the natural language and the machine language is greater for non-English-speaking students than for their English-speaking colleagues. While translating this task into TI-89 language, many students forget to add the comma and the x. They do not see the need for the extra letter at the end. The next task in the same assignment creates some confusion:

$$\text{What happens if you change 3 into 3.0?}$$

Some students enter solve(x^2 = 3, 0). They mistake the comma in the syntax of the solve command with the decimal comma that is used in the Netherlands as the alternative for the decimal point.

Algebra on Screen, on Paper, and in the Mind

Things turn out better in assignments 13 and 14 (see **fig. 14.1**), where the equation has more variables.

> **13** Enter: solve(a * x + b = 5, x)
>
> *Note: an * needs to be inserted between a and x; otherwise ax will be considered as a word variable. The answer of solve can be a formula: x is expressed in a and b.*
>
> **14** Change the last x in the solve command of the previous exercise into b. Now the equation is solved with respect to b and b is expressed in a and x.

Figure 14.1. Solving an equation with three variables

One of the students, Barbara, has solved the equation correctly with respect to x and with respect to b. She copies the results $x = -(b-5)/a$ and $b = 5 - a*x$ into her notebook. Then the observer passes by, and the following dialogue takes place.

> *Obs:* Do you think this [the answer of 13] is a satisfying answer?
>
> *B:* …Yeah, you have to get out the x and look how the formula gets if you want to know the x.
>
> *Obs:* Yes. And what is the difference between 13 and 14?
>
> *B:* There [points at 14] you want to know the b.
>
> *Obs:* Yes. And if you want to know the a, what is the command that you have to enter?
>
> *B:* With an a after the comma.

Barbara sees that x is the variable she wants to know and that after the comma she needs to specify the "unknown-to-be-found" as defined by Bills (2001). Barbara's attention seems to be focused on the syntax of the command and the necessity to specify a variable rather than the fact that the solution is an algebraic expression. The dialogue with Barbara is representative of reactions from other students, such as Marg, who seems to have a more practical style: do what you are asked to do and do not wonder too much.

> *Obs:* You found the right solution for 13 I see, and for 14 as well. Don't you think it is strange, such a long story as a solution?
>
> *M:* No, because they ask [reads once more the assignment], they ask, in fact they only say that there has to be such a

thing in between [a comma?], I don't believe it concerns a real problem.

Obs: No, OK, that's true. But you had that equation, $a * x + b = 5$, and now in exercise 13 I get $x = -(b-5)/a$.

M: Yeah because look you had to calculate x, so you had to know what x was.

Obs: And what is the difference, how is that in assignment 14?

M: Well, in 14 you need to know what b is, there you have to know how to calculate b.

Obs: Yes. And if you would enter solve something comma a?

M: Well, then you get $a = \ldots$ a story.

Another student, Suzanne, looks further than the syntax of the solve command. She correctly explains why any expression can be the outcome of a solution process. This type of awareness of the outcome of the command was uncommon in the experimental classes. Like Barbara, Suzanne wrote in her notebook: $x = -(b-5)/a$ and $b = 5 - a * x$.

Obs: Is that $[-(b-5)/a]$ really an answer?

S: Yes.

Obs: Why?

S: Well, you can give anything as an answer, I mean you could also say that for example $5a$ is not good, I mean it probably is not the smallest that you can get.

Obs: How do you mean, the smallest?

S: Well maybe you can simplify it.

Obs: I see. Would that be possible here, simplify?

S: Let me see, well I don't think so, because for ... $b - 5$ is a thing, so if you don't know what b is, you cannot write it easier.

Later in the same lesson, one of the students, Mandy, is invited to show her results to the class using the viewscreen of the calculator. Mandy enters $solve(a * x + b = 5, x)$ and gets the right answer, of course. Next, the teacher starts a discussion and Cedric helps Mandy. Later Ada enters the discussion.

Teacher: Why do you do comma x?

M: Because it is in the textbook?

T: But do you know why?

M: Well, not really, in fact. Yeah, because you have to calculate x apart, or something.

C: Because you want to know the x.

M: Yeah, that.

T: Now I want to have an answer with $b = $.

M: Then you have to change the x into a b, because you want to know that one, then you know b. [M edits the first command and gets the solution for b.]

T: Could you solve another letter?

M: a. [M finds the solution for a.]

A: The a equals the x!

T: Is the a equal to the x?

M: Yeah, well, they are turned around.

[The students see that the formula is symmetric in a and x, so in the solutions for a and x, the letters are interchanged.]

Again, we see that the comma-x is perceived as an indication of the letter that you want to put apart. Ada's remark shows how only a few students really look at the answers of the computer algebra device and are able to take these answers as a starting point for reflection.

How do we link the student episodes described in the foregoing with the theoretical perspectives? The question of the required "comma-letter" in the syntax of the solve command can be discussed in terms of instrumentation. In the technical aspect of the solve instrumentation scheme, the necessity to add "comma-letter" to the equation is difficult. This technical obstacle has a mental counterpart: if students are unaware that an equation is always solved *with respect* to a certain variable, they will not understand this requirement and will tend to forget it. However, the fact that the TI-89 requires the letter specification may be helpful in making explicit the idea of "solving with respect to." The words of Barbara ("you have to get out the x"), the words of Marg ("you had to know what x was"), and the words of Mandy ("you have to calculate x apart") illustrate this principle. We see the relationship between computer algebra technique and mental conception. When using paper and pencil, planning to solve an equation and carrying out the solution procedure often take place simultaneously, thus obscuring the implicit choice of the letter with respect to which they solve.

The question of accepting an expression as the outcome of the solution process can be viewed as a process-object matter. The dialogue with Suzanne indicates that she looks at the outcome and accepts an expression as a *result*, seeing the expression as an *object* and clearly expressing this perception in the phrase "$b - 5$ is a thing." Also, Ada's reaction indicates that she actually looks at the formulas and compares them so as to discover patterns. However, many other students do not seem to look too carefully at the answer when it is an expression. When they do reflect on the answer, their understanding is often dominated by the procedural view: the solution indicates how "to calculate x apart" (Mandy) "so you had to know what x was" (Marg). We have the impression that these students view the expression as a recipe, a description of the process of calculating the concrete numerical value of the unknown-to-be-found—a process rather than an object that can be the answer to the solution procedure. The process aspect dominates the conception of an expression. We conjecture that this perception makes the goal of taking full advantage of the power of computer algebra more difficult to attain.

The Rectangular Triangle

In the previous section, we saw that many students did not care too much about an expression as a solution of an equation. In the geometrical context of the rectangular triangle assignment (see **fig. 14.2**), two equations with two variables are used. The solutions cannot be ignored any longer. This example is described in more detail in Drijvers and Van Herwaarden (2000).

For this assignment, the textbook proposes a solution scheme called Isolate-Substitute-Solve. In the actual situation of question **a** in **figure 14.2**, the equations are $x + y = 31$ and $x^2 + y^2 = 25^2$, where x and y stand for the lengths of the two perpendicular edges. **Figure 14.3** shows the proposed solution strategy on the TI-89.

First, one of the variables, in this instance, y, is isolated in one of the equations. The result is then substituted into the other equation using the "where-in-operator," symbolized by the vertical bar $|$. In the resulting equation, only one variable is left (in this example, x) and a second solve-command yields the answer, $x = 7$ or $x = 24$. The calculation of the corresponding values of y is no longer difficult.

This step-by-step procedure corresponds to a problem-solving strategy that could be carried out with paper and pencil. However, the first step is not trivial. Many students have conceptual difficulties in using the command "solve" to isolate one of the two variables. To many students "solve" means finding a solution, whereas $31 - x$ in this context is an expression, not a

Algebra on Screen, on Paper, and in the Mind

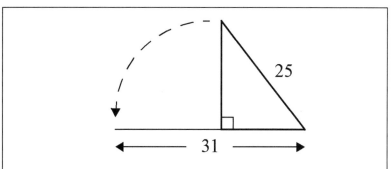

The two right-angle edges of a rectangular triangle together have a length of 31 units. The hypotenuse is 25 units long.

a. How long is each of the right-angle edges?

b. Solve the same problem when the total length of the two edges is 35 instead of 31.

c. Solve the problem in general, without the given values of 31 and 25.

Figure 14.2. The rectangular triangle assignment

solution for y. A solution in this situation apparently should be a concrete number, preferably a decimal number. Maybe the concrete geometrical context leads students to expect concrete numerical results more strongly than the context-free problems in the previous section does. Whatever the reason, the fact that the same solve command is used on the TI-89 for numerical solutions and for the isolation of a variable requires an extended conception of solve: it also stands for "taking apart a variable" and for "expressing one of the variables in terms of one or more others in order to process it further."

Figure 14.3. The solution scheme on the TI-89

Some of the students who do understand the use of solve for obtaining an intermediate result use this tactic twice, as the following dialogue with Martin shows.

> *M:* This x, I have to say both made y out of it by doing solve.
>
> *Obs:* Yes, both solve with respect to y.
>
> *M:* How do you call that, substitute or so?
>
> *Obs:* I call it isolate ...
>
> *M:* Isolate.
>
> *Obs:* ...because you put the y apart.
>
> *M:* OK. And then the other one as well, and I put that against each other and then you get .... [M continues with solving the equation $31 - x = \sqrt{(625 - x^2)}$.]

Martin's approach is close to the solution procedures that the students previously used by hand for calculating the intersection point of two lines. In this way Martin creates congruence between the paper-and-pencil method and the machine technique. Furthermore, this strategy avoids substitution, the topic of the next section of this chapter.

The generalization of the problem in part **c** of the assignment (see **fig. 14.2**) presents other difficulties. Some students seem to be less willing to accept parameters with unknown values than they were in the abstract situations presented in the preceding section. They want to know its numerical value before they are able to proceed. Küchemann (1981) called this phenomenon *letter evaluation*. For example, Anne and Maria worked out questions **a** and **b** but have difficulties with question **c** (see **fig. 14.2**). They ask the observer for assistance.

> *Obs:* The 31 first changed into 35, and now it changes again into something else.
>
> *A:* In something.
>
> *Obs:* Pardon?
>
> *A:* In simply, that you don't know.
>
> *Obs:* Yes. Well let's take a letter for it.
>
> *A:* Oh, but then you cannot calculate it.
>
> *M:* s.

Obs: s, for example. Instead of that 35, I now fill in an *s*, and that has to be the square of 25; oh no, you don't know the 25 either.

A: Exactly!

Obs: Well let's take another letter for that.

M: v.

[Then the system of equations is solved, but Anne does not seem to be convinced.]

Looking back at the observations in this section in the light of the theoretical framework, we see the following outcomes. If the equation contains more than one variable, the mental conception of solve has to be extended to isolating and expressing one variable in terms of the others before the student is able to see that solve is the correct command to use for isolation. This extension is probably more difficult if the outcome of the isolation is an intermediate result than if it is the final outcome of an exercise. These findings illustrate the interaction between machine technique and conceptual development in the instrumentation scheme: the fact that the same command is used for isolating a variable as for calculating numerical results indicates that these operations—quite different in the eyes of the student—are identical in the world of computer algebra and are mathematically equivalent.

Martin's approach shows that the user naturally tries adapting the existing paper-and-pencil technique to the computer algebra instrument, thus creating congruence between the two media and the accompanying ways of working.

Finally, some students tend to see numbers as favorable outcomes and perceive a letter as a reference to such a number. The letter invites an evaluation process; it is not an object on its own. Anne's comment "Oh, but then you cannot calculate it" illustrates this point. Related to this response is the idea that the concrete context makes more difficult students' tasks of accepting an expression as an object that is an answer and of seeing the meaning of general parametric solutions.

The TI-89 "Wherein-Operator" | and the Conception of Substitution

As a second example of the interaction among mental mathematical conception, paper-and-pencil work, and computer algebra use, I address the substitution of expressions that contain parameters. In the previous section we saw that students had already formed a mental conception of what solving an equation means. This conception was developed by paper-and-pencil work and was extended and adapted to computer algebra use. In substitution, a similar development took place; however, the students seemed to have very limited preconceived conceptions. Their idea of substitution was restricted to the experience of replacing variables by numerical values. The use of computer algebra stimulated their building of an extended conception, at least as far as the performance of the procedure is concerned.

The Volumes of a Cylinder and a Cone

In the introductory part of the textbook, the students first substitute some concrete values for a variable x. For example, $3x + 5 \mid x = 2$ should be read "$3x + 5$ with $x = 2$" or "$3x + 5$ wherein $x = 2$." They then replace x by another variable, which is new to them, and by the expression "$a + 1$." Then the cylinder assignment (see **fig. 14.4**) is given to the students. The last line in the assignment serves as a visualization and an opportunity for self-control.

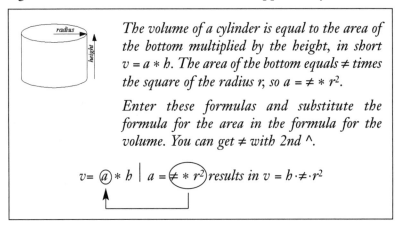

Figure 14.4. The cylinder assignment

Only a few students have difficulty with this task. The observer intervenes and asks for the meaning of the vertical bar, the $\mid$ symbol. Let us look at the answers of Barbara and Tony in the dialogues that follow. Barbara formulates the idea of substitution as a replacement action quite clearly; for Tony, this concept seems to be less clear. His perception of substitution merely reflects the combining of two formulas.

Obs: What does that mean, that vertical bar, what does it do?

B: Then you can [fill in] one of the letters of the formula ..., there you can fill in what number has to be there.

Obs: Fill in what number ..., but here [in the assignment] there is no number.

B: No, there is a formula to calculate the *a* with.

Obs: And what happens to that formula?

B: Which one of the two do you mean?

Obs: The one behind the bar?

B: Well, in fact, it is filled in in the place of *a*.

The second dialogue proceeded as follows:

Obs: Now what exactly does that vertical bar mean?

T: It means that the left formula is separated from the right, and that they can be put together.

Obs: And what do you mean by putting together?

T: That if you, that you can make one formula out of the two.

Obs: How do you do that, then?

T: Ehm, then you enter these things [the two formulas] with a bar and then it makes automatically one formula out of it.

From the dialogue above and similar observations, we conclude that many of the students perceive the substitution, symbolized by the vertical "wherein-bar," as an action of replacement. This replacement is not limited to numbers but can also be applied to expressions. The drawing in the last line of the assignment seems to be an important visualization. The TI-89 notation is close to the paper-and-pencil method of substituting, and it also corresponds to the mental image of substitution as replacement.

The written test given at the end of the experiment contained a similar assignment for one of the classes (see **fig. 14.5**).

Part **a** (see **fig. 14.5**) was easily solved by most of the students. However, only three out of twenty-seven students received full credits for part **b**, and twelve out of twenty-seven students received a partial score. The remaining twelve students received no score at all. The partial scores were the most interesting. Often the answer was correct but not given in a completely simplified form or contained a minor error, such as the following:

1. $\frac{1}{3}\pi r^2 * (2r)$

2. $\frac{1}{3}\pi \rho^2 * r^2$

3. $\frac{1}{3} * \text{area} * (\text{radius} + \text{radius})$

4. $\frac{1}{3}\pi x^2 * y \mid y = 2x$

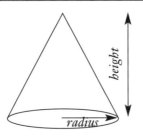

*The volume of the cone is equal to the area of the bottom multiplied by the height divided by 3, so volume = ⅓ * area * height.*

*The area of the bottom equals ≠ times the square of the radius r, so area = ≠ * radius²*

a. *Calculate the volume of the cone if the radius is 5 and the height is 9.*

b. *Suppose you do not know the height and the radius, but you do know that the height is equal to the double radius. What formula for the volume can you infer in that case?*

Figure 14.5. The cone assignment

The answer shown in item 1 above is correct. Interestingly, the student did not use the TI-89 to get this result, because the machine would have simplified the expression to $⅔ ≠ r^3$. Apparently, this outcome was obtained mentally or by using paper and pencil. The same assumption holds for the answer in item 2 above: the student probably wrote r^2 instead of $2r$ but also chose not to use the TI-89 for the evaluation. The answer in item 3 above reveals a similar way of working. In item 3, the formula for the area was not substituted, and similarly the expression "radius + radius" was not evaluated, so it was obtained without the machine. The answer in item 4 is correct in TI-89 notation, but the result is not written down on paper. Was this substitution executed on the machine? In item 5, only the expression for the area is substituted. The relation $h = 2*r$ is not substituted, and the TI-89 was not used.

The two examples in this section show that the development of the students' understanding of substitution seems to be supported by the notation of the wherein-bar of the TI-89. The machine technique seems to stimulate the development of the conception of substitution as a replacement mechanism. That technique is typically instrumentation; the machine technique and the conceptual development influence each other in a reciprocal way. As far as the process-object duality is concerned, many students seem to accept the substitution of nonnumerical values, although the result of the cone assignment is somewhat disappointing. I hope that the operation on an expression by means of substituting it provokes a further reification of such an expression. However, the explanations of Barbara ("there you can fill in what number has to be there" and "there is a formula to calculate the *a* with") show that the operational aspect in the conception of the expression is still too important to her.

We were surprised that so many students preferred to carry out the substitution mentally or with paper and pencil, without using the computer algebra device. We see two explanations for this phenomenon. First, some students may have felt uncertain about their machine skills in the stress of the assessment situation. Second, they may simply have found it easier to do the substitution by hand. In either situation, the students transferred their machine technique to a paper-and-pencil technique, an interesting achievement. However, in a more complex problem-solving strategy, the substitution raises more difficulties, as discussed in the next section.

The Rectangular Triangle Revisited

In a preceding section, I presented the problem of the rectangular triangle. The situation described in part **a** of **figure 14.2** led to the equations $x + y = 31$ and $x^2 + y^2 = 25^2$. We described how the problem-solving strategy started with isolating one of the variables in at least one of the equations. For the students, this means of using "solve" was new. The next step, substituting the isolated expression into the other equation, is the most difficult point. In spite of the growing conceptual insight that we meant to see in the preceding section, substitution is often an obstacle in this embedded situation. The substitution is tangled up between two solve commands (see **fig. 14.3**), the first command for the isolation of a variable and the last command for the calculation of the final solution. Furthermore, students' conception of substitution may rely too much on the mechanism and not enough on its meaning.

Three kinds of difficulties regularly occur. First, students may try to substitute the isolated form into the equation from which it has been derived (see **fig. 14.6**, line 1). Of course, the message "true" causes some confusion, and often students do not understand the logic behind it. For example, during a classical demonstration with the viewscreen, Sandra enters

$$\text{solve}(x + y = s \mid y = s - x, x).$$

The machine replies "true." Rob adequately comments, "She did not use the second equation."

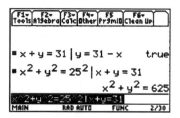

Figure 14.6. *Difficulties with the substitution*

A second frequent and persistent mistake is the substitution of a non-isolated form (see **fig. 14.6**, line 2.) In set theory, of course, the notation $\{x^2 + y^2 = 25^2 \mid x + y = 31\}$ would be correct. However, in this situation the notation reveals an incomplete understanding of the substitution mechanism. In the following fragment, John explains this issue adequately to Rob, but such explanations were rare.

[Rob wrote down "solve($o^2 + a^2 = 625 \mid o + a = 31, a$)."]

J: You have to put one apart, you have to put either the *o* or the *a* apart.

R: You can only work on one letter?

J: I think, if you have got the vertical bar, you are allowed to explain only one letter, so *o* =.

The third common mistake involves the combination of solve and substitution. As Sandra's input above shows, some students try to combine the substitution and the solution in one line:

$$\text{solve}(x^2 + y^2 = 25^2 \mid y = 31 - x, x)$$

The problem with this nested form is the choice of the letter at the end. With respect to which letter do we need to solve? Several times, students chose the wrong letter, the *y* instead of the *x*, after the comma. In such instances, the substitution is performed. However, because the letter *y* does not appear in the resulting equation, it cannot be solved with respect to *y*; therefore, it is returned in an equivalent form. To avoid this error, students need to mentally split the nested command into two subprocesses. They need to understand that if the substitution part starts with "*y* =", then the resulting equation will no longer contain *y*; therefore, it must be solved with respect to *x*.

This problem-solving behavior shows that students' insight into the substitution process is not mature enough to handle this more complicated situation satisfactorily. One of the students avoided this mistake by symbolizing the two subprocesses in the command with an extra pair of brackets:

$$\text{solve}((x^2 + y^2 = 25^2 \mid y = 31 - x), x)$$

In summary, we see that in more complicated situations, students have difficulties keeping in mind the correct idea of substitution. More specifically, students regularly try to substitute an isolated form into the equation from which it originated, or they try to substitute a nonisolated form. I again perceive these errors as matters of instrumentation. If students have an appropriate mental image of substitution, they should be able to perform the substitutions in the computer algebra environment correctly. Classroom discussions on these instrumentation problems can improve students' understanding of substitution and strengthen the machine skills needed to solve problems like those shown above.

When a student substitutes $y = 31 - x$ or $y = s - x$, the perception of the expression as a calculation recipe for concrete values may be an obstacle. Also, the wish to evaluate s can hinder the substitution. If the input of the substitution is an expression, however, its object character may be reinforced by the process of substitution. Thus, substitution can support the reification of an expression.

Students who use the nested command that combines solve and substitute must be aware of the different role that each letter plays at each point in the process. This mental skill is reflected in the technical skill to choose the final letter in the command correctly. The student who used extra brackets in solve$((x^2 + y^2 = 25^2 \mid y = 31 - x), x)$ found a means to express the conceptual insight in the technical notation—an example of good instrumentation.

Substitution of a Parametric Equation

As I mentioned previously, the educational experiments started with a pretest. One of the assignments of the 2001 experiment pretest is shown in **figure 14.7**. In the pretest were six good answers, nine incorrect answers, and thirty-six instances in which students did not answer the question. Among the nine incorrect responses, strange answers, such as $x^4 = 10$ or $x^2 + 3^2 = 10$, surfaced, but a few "nearly good" answers, such as $x^2 + a - x = 10$, also appeared. The global impression is that a great majority of the students simply do not know what to do with this question.

> *Consider the equations* $y = a - x$
>
> $$x^2 + y^2 = 10$$
>
> *Make one equation from these in which y does not appear.*
> *You do not need to solve this new equation!*

Figure 14.7. Pretest and post interview assignment

After the experiment, interviews were held with nineteen of the students. The assignment shown in **figure 14.7** was one of the questions asked again during the interview. Because the students had already handed in the TI-89, they were forced to use the traditional paper-and-pencil approach. Of these nineteen students, fourteen found the correct answer, $x^2 + (a - x)^2 = 10$, or, in one instance, $a - x = \sqrt{(10 - x^2)}$. Some of these fourteen students first made an error that they later corrected after receiving a hint from the interviewer. Four other students performed the substitution but made an error that was not corrected by the interviewer. The answers of the latter four students are shown below:

1. $x^2 + a^2 - x^2 = 10$
2. $x^2 + a - x = 10$
3. $a - x = 10 - x^2$
4. $a - x = \sqrt{(10 * x)}$

In answer 1 above, the student squared *a* and squared *x* instead of squaring their difference. In answer 2 above, the student forgot the squaring. In answer 3, the student missed the square root on the right-hand side. In answer 4, the student multiplied 10 and *x* while isolating *y* instead of subtracting x^2 from 10. However, in spite of the errors in the performance, all four students showed some understanding of the principle of substitution. Only one student out of nineteen did not arrive at a sensible answer.

The following fragment of the interview with Donald is illustrative. After some struggle with the task, Donald carries out the substitution correctly and gives a nice explanation.

> *Int:* You understand the problem, don't you?
>
> *D:* Yes, I don't know how to do it without calculator.
>
> *Int:* How would you do it, then, with your calculator?

D: Well, eh, solve, and then, let me think, no, eh, yes, I think solve doesn't contain y,

Int: Ehm…

D: Gosh, eh, maybe no, wait a moment, let me have a look. [D writes down $x^2 + (a - x)^2 = 10$.]

Int: Wonderful, how do you get this?

D: I was thinking too difficultly. Here is already a y, and here it says what y is, so you just have to replace this with that; I was just thinking too difficultly.

Int: Right. And do you know now how you would do it with the calculator? Of course, you could type it in like this.

D: Then I would do this… with $y = a - x$. [D adds to $x^2 + y^2 = 10$ at the sheet of paper: $|\ y = a - x$.]

Of course, the postexperiment interviews cannot be compared with the written pretest. The interviewer was able to rephrase the question and offer comments relating to the task. Nevertheless, almost all the students performed a substitution of a parametric expression for y without a computer algebra machine, even if they did not do so correctly in all instances. We interpreted this outcome as modest evidence that working on the substitution in the computer algebra environment improved their mental conception and also their paper-and-pencil skills.

The transfer from computer algebra technique to paper-and-pencil work also appears in the students' notation and language, which in some cases explicitly referred to the TI-89 machine. For example, Donald wrote down the vertical bar notation for substitution. Another example of this transfer of knowledge is the way Kevin looks at the equation $y = a - x$. In his reasoning he makes use of TI-89 language, but he also relates the machine technique to his understanding. Remarkable, by the way, is the fact that he converts the first isolated equation to the symmetrical, nonisolated form.

K: Well, I think $x + y = a$, isn't it?

Int: Yes.

K: But in fact this is, this is already solved. Because if you do solve, $x + y = a$, comma y with solve, then you get this: $[y = a - x]$.

Another student, Martin, uses the strategy of double isolation, as we saw already in a preceding section. By mistake, he isolates y in the second equation by hand as $y = \sqrt{(x*10)}$. His equation then is $a - x = \sqrt{(x*10)}$, but he does not feel sure about the right-hand side. Next, he shows how he could solve the problem with the TI-89, using the appropriate commands.

> *Int:* And how would you do that if you had the calculator in your hands?
>
> *M:* Then I would have done solve.
>
> *Int:* For the second equation?
>
> *M:* I would have done solve for the second [the quadratic equation]. And then you can put solve in front of this one as well. [On paper he changes $x^2 + y^2 = 10$ into solve $(x^2 + y^2 = 10, y)$ and $a - x = \sqrt{(x*10)}$ into solve$(a - x = \sqrt{(x*10)}, x)$.]

Looking back at the assignment in this section and its results, I have three comments. First, the instrumentation scheme for substitution in the computer algebra environment stimulated the concept building for the students: their conception was extended to the substitution of expressions. Second, transfer to the paper-and-pencil situation apparently took place as a result: most of the students performed the substitution of a parametric expression by hand after the experiment, and some of them explicitly referred to the TI-89 environment using language and notation. This transfer is possible because the TI-89 substitution technique and the paper-and-pencil procedure are quite congruent. Third, some students seem to perceive expressions as objects and not only as processes. Donald's comments as he refers to the expression as a *thing* and not as a calculation recipe ("here it says what y is" and "you just have to replace this by that") offer some evidence for this conclusion.

Discussion

We presented examples of students' behavior in assignments where the operations solve and substitute play a role, either performed with the TI-89, mentally, or on paper. We were interested in the relationship among computer algebra use, paper-and-pencil work, and algebraic thinking. We tried to interpret the observations by using the theoretical perspectives of the process-object duality and the instrumentation issue. In this section we discuss some remaining questions:

1. What is the role of computer algebra in the process-object development?
2. How conceptual is the instrumentation, and what is the role of the context?
3. Is a difference evident between the case of solve and the case of substitute?
4. What are the consequences for teaching?

To answer question 1 above, we summarize the relation between the process-object issue and the episodes of this chapter. On the one hand, a limited mental concept of an expression as a calculation process clearly can hinder its substitution in the computer algebra environment and its acceptance as a solution of an equation. On the other hand, the substitution of an expression in the computer algebra environment can provoke the development of its object aspect in the mental conception and reinforce reification. In general, the computer algebra environment tends to reduce processes too complex to carry out by hand to one single command. As a result, the process becomes more static and will get the character of an object to a greater extent than if performed by hand. For example, Dubinsky (1991) suggests that performing processes using a computer may stimulate its encapsulation. As Monaghan (1994) points out, the mathematical entities in such an environment may tend to have a structural character, whereas the processes are more distant from the objects than while working in a paper-and-pencil mode. Also, a flexible way of dealing with literal symbols as objects instead of pseudonumbers is required.

More recently, the role of computer algebra in the object-process duality is described as follows by Heid (2001):

> Computer algebra systems seem to be ideal technologies for fostering growth in the level of understanding of mathematical concepts. Using the CAS, students can signify (and execute) actions on mathematical entities without needing to carry out the procedures by hand. A CAS that allows the user to craft his or her own function rules and small programs provides a tool for capturing processes so that actions can be performed on them. I am not claiming that because a student writes a routine to perform a procedure that he or she can operate on that procedure as an object, but rather that the availability of this accurate and convenient tool for condensing a process into a single command has potential as an intermediate step between process and object understandings.

To answer question 2 above, we conclude that the instrumentation of solve requires a technical mastering of the syntax of the TI-89 command and a mental adaptation of the solve concept. In a similar way, the instrumentation of the substitution involves the mental development of the substitution of expressions. However, the question is how "deep" the conceptual developments were. Many errors were made, and some of them were quite persistent. I wonder whether the students who substitute an expression into a formula, or who solve a parametric equation with the computer algebra device, perceive the expressions as mathematical objects or merely as collections of symbols and letters. Maybe the initial acceptance of expressions as results from the solve command has to do with a limited meaning attached to the results. As soon as the expressions acquired meaning, the instrumentation seemed to be more problematic, maybe because of the concrete contextual references. In the final assignment, this contextual reference is absent again. Not clear is whether the students who arrived at solving this problem have just a procedural knowledge or if this knowledge is embedded in a richer conceptual framework. Also, the shift between contextual reference and abstraction—of utmost importance in algebra—requires attention.

In answer to question 3 above, if we compare the observations concerning solve with the observations on substitution, we see a difference. The transfer from computer algebra technique to paper-and-pencil procedure that we noticed for substitution was not observed for solving equations. A possible explanation for this absence is that the solve procedure in the computer algebra environment is a black box to the students. The way the process is carried out inside the machine is obscured, so the relationship to the paper-and-pencil methods is lost. The students have no insight into the way the TI-89 internally solves an equation; therefore, they have difficulty relating the paper-and-pencil work to the machine output. This situation is different for substitution. The TI-89 notation for substitution is a natural one, and in simple cases the student can easily verify what is happening. Therefore, the computer algebra approach is congruent to the paper-and-pencil approach, explaining why transfer can take place.

Finally, to answer question 4 above, we ask, What are the consequences of the observations for the teaching? For a teacher who uses any IT tool in the mathematics teaching, paying explicit attention to the development of instrumentation schemes can be worthwhile. While teaching with IT tools, a recommended approach is to take care of a simultaneous development of both the technical and the mental aspect of the instrumentation schemes and to relate these aspects to existing paper-and-pencil techniques. Also, the concept of instrumentation can be valuable while trying to understand the difficulties met by the students. In fact, many difficulties that students encounter while working with a computer algebra device can be viewed as instrumentation obstacles. These instrumentation roadblocks are often interesting to consider because they may reveal an incomplete conceptual part of the instrumentation scheme. Keeping in mind the two sides of the instrumentation scheme—the technical and the mental part—can be useful in understanding students' difficulties and helping to overcome them.

Conclusion

In this chapter we saw that mental conceptions, computer algebra techniques, and paper-and-pencil methods are closely related to one another. Mental conceptions can be influenced by machine techniques. Students can introduce machine notations in their paper-and-pencil work. Difficulties in computer algebra techniques can be attributed to limited conceptual understanding. Also, the work in the computer algebra environment sometimes seems to lead to better mental or paper-and-pencil problem-solving skills.

To establish the complex relationship among machine technique, mental conception, and paper-and-pencil work, students must perceive the congruence among them. Students should be able to "match" the mental image, the traditional way of problem solving, and the approach in the computer algebra environment. In the meantime, the different media may complement each other and can stimulate the development of the other forms. We should try to use the mental conceptions, the computer algebra device, and the paper-and-pencil work to support one another and to coincide into an integrated conception. A practical strategy for teachers is to ask students who are working in a computer algebra environment how they would proceed if they had only paper-and-pencil, and vice versa.

The two theoretical perspectives can be valuable in interpreting students' behavior and in helping students overcome obstacles while working in a computer algebra environment. The process-object issue shows that the computer algebra environment tends to emphasize the object side of mathematical concepts, often not the dominating aspect for the students. While teaching, we can be helped by keeping this duality in mind and by ensuring that the process aspect also receives attention. The theory of instrumentation shows how the mental conception and the machine technique are interwoven; thus, it is an adequate framework to better understand students' difficulties. In fact, an error that occurs while using a machine may reveal a lack of congruence between machine technique and mathematical conception or may indicate limitations in the conceptualization of the mathematics involved.

REFERENCES

Artigue, Michèle. "Rapports entre dimensions technique et conceptuelle dans l'activité mathématique avec des systèmes de mathématiques symboliques." In *Actes de l'université d'été* (1997): 19–40. Rennes: IREM de Rennes, 1997.

Bills, Liz. "Shifts in the Meanings of Literal Symbols." In *Proceedings of the 25th Conference of the International Group for the Psychology of Mathematics Education*, edited by M. Van den Heuvel-Panhuizen, pp. 2-161–2-168. Utrecht, The Netherlands: Freudenthal Institute, 2001

Drijvers, Paul. "Students Encountering Obstacles Using a CAS." *International Journal of Computers for Mathematical Learning* 5(3) (September 2000a): 189–209.

———. *Introductie TI-89*. Utrecht: Freudenthal Instituut, 2000b.

———. *Veranderlijke Algebra*. Utrecht: Freudenthal Instituut, 2000c.

———. "The Concept of Parameter in a Computer Algebra Environment." In *Proceedings of the 25th Conference of the International Group for the Psychology of Mathematics Education*, edited by M. Van den Heuvel-Panhuizen, pp. 2-385–2-392. Utrecht, The Netherlands: Freudenthal Institute, 2001.

Drijvers, Paul, and Otto van Herwaarden. "Instrumentation of ICT-tools: The Case of Algebra in a Computer Algebra Environment." *International Journal of Computer Algebra in Mathematics Education* 7(4) (2000): 255–75.

Dubinsky, Ed. "Reflective Abstraction in Advanced Mathematical Thinking." In *Advanced Mathematical Thinking*, edited by D. Tall, pp. 95–123. Dordrecht: Kluwer Academic Publishers, 1991.

Gray, Eddie M., and David O. Tall. "Duality, Ambiguity, and Flexibility: A 'Proceptual' View of Simple Arithmetic." *Journal for Research in Mathematics Education* 25(2) (March 1994): 116–40.

Guin, Dominique, and Luc Trouche. "The Complex Process of Converting Tools into Mathematical Instruments: The Case of Calculators." *International Journal of Computers for Mathematical Learning* 3(3) (1999): 195–227.

Heck, Ardie. "Variables in Computer Algebra, Mathematics and Science." *International Journal of Computer Algebra in Mathematics Education* 8(3) (October 2001): 195–221.

Heid, M. Kathleen. "Theories That Inform the Use of CAS in the Teaching and Learning of Mathematics." http://ltsn.mathstore.ac.uk/came/events/freudenthal/3-Presentation-Heid.pdf. (23 August 2001).

Küchemann, D.E. "Algebra." In *Children's Understanding of Mathematics: 11–16*, edited by K. M. Hart, pp, 102–19. London: John Murray, 1981.

Lagrange, Jean-Baptiste. "Learning Pre-Calculus with Complex Calculators: Mediation and Instrumental Genesis." In *Proceedings of the XXIIIrd Conference of the International Group for the Psychology of Mathematics Education, Vol 3*, edited by O. Zaslavsky, pp. 193–200. Haifa, Israel: Technion, 1999a.

———. "Complex Calculators in The Classroom: Theoretical and Practical Reflections on Teaching Pre-Calculus." *International Journal of Computers for Mathematical Learning* 4 (1999b): 51–81.

———. "Techniques and Concepts in Pre-Calculus Using CAS: A Two-Year Classroom Experiment with the TI-92." *International Journal of Computer Algebra in Mathematics Education* 6(2) (1999c): 143–65.

Lagrange, Jean-Baptiste, Michèle Artigue, Colette Laborde, and L. Trouche. "A Meta Study on IC Technologies in Education." In *Proceedings of the 25th Conference of the International Group for the Psychology of Mathematics Education*, edited by M. Van den Heuvel-Panhuizen, pp. 1-111–1-122, Utrecht, The Netherlands: Freudenthal Institute, 2001.

Monaghan, John. "On the Successful Use of Derive." *The International Derive Journal* 1(1) (April 1994): 57–69.

Rabardel, Pierre. *Les Hommes et les Technologies—Approche Cognitive des Instruments Contem-porains*. Paris: Armand Colin, 1995.

Sfard, Anna. "On the Dual Nature of Mathematical Conceptions: Reflections on Processes and Objects as Different Sides of the Same Coin." *Educational Studies in Mathematics* 22 (1991): 1–36.

Tall, David O. "Mathematical Processes and Symbols in the Mind." In *Symbolic Computation in Undergraduate Mathematics Education*, MAA Notes 24, edited by Z. A. Karian. Washington, D.C.: Mathematical Association of America, 1992.

Tall, David, and Michael Thomas. "Encouraging Versatile Thinking in Algebra Using the Computer." *Educational Studies in Mathematics* 22 (1991): 125–47.

Trouche, Luc. "La Parabole du Gaucher et la Casserole a Bec Verseur: Étude des Processus D'apprentissage dans un Environnement de Calculatrices Symboliques." *Educational Studies in Mathematics* 41 (March 2000): 239–64.

Activity 8

Use the sequence operation to produce the sequence 3, 6, 9, 12, 15 as many different ways as you can.

Solution:

- seq(n,n,3,15,3)   {3 6 9 12 15}
- seq(3·n,n,1,5,1)   {3 6 9 12 15}
- seq(n+3,n,0,12,3)   {3 6 9 12 15}
- seq$\left(\left(\sqrt{3\cdot n}\right)^2,n,1,5,1\right)$ {3 6 9 12 15}
- seq(20−n,n,17,5,⁻3)   {3 6 9 12 15}

Will this instruction work?   seq(n∗cos(2nπ),n,3,15,3)

CHAPTER 15

Learning Techniques and Concepts Using CAS: A Practical and Theoretical Reflection

Jean-Baptiste Lagrange

RESEARCH STUDY relating to the use of CAS in secondary school teaching and learning has existed in France since 1994, connecting teachers who are developing innovative classroom uses of computers with the international community of researchers in this domain. A team of educators—myself included—lead by Michele Artigue of Universitè Paris VII became more and more concerned about the emphasis that teachers and researchers placed on the conceptual dimension of using CAS to teach and learn mathematics, thus possibly neglecting the role of other dimensions. Studying the use of Derive in French classrooms from 1994 to 1996, we observed that teachers often approached the use of CAS by neglecting the role of the technical work in students' learning. The role of technical work—both with and without CAS—and how it is linked with students' understanding of mathematics became considerations.

Since CAS became available on the main window of such calculators as the TI-92, daily classroom use of CAS became possible. In 1996, in response to more widespread CAS use, we started a new study to assess the relevance of the TI-92 symbolic calculator for learning precalculus in the science "stream" of French schools.

In this chapter, I report how our team of educators recognized a link between the technical work in learning mathematics and the understanding of mathematics. Then I show how we conceptualized this link between the two components, how this conceptualization helped us design classroom sessions, and what perspectives it opens.

The Role of Techniques Using CAS

In teaching mathematics—especially algebra—the tension between manipulation and understanding is ancient. Rachlin (1989) states, "Teachers [in the USA] even [in 1890] were opposed to what they saw as an overemphasis on manipulative skills and were calling for a meaningful treatment of algebra that would bring about more understanding." In the past fifteen years, the idea that universal access to the new technology would "enable us to modify our skill-dominated conception of school algebra and rebalance it in favor of objectives related to understanding and problem solving" (Kieran and Wagner 1989, p. 8) gained greater acceptance. Some people had reservations about this agenda and expressed concerns that universal access would require "the mathematics education community [to] deal(s) with the four factors of teachers, textbooks, evaluation and articulation between . . . coursework" (Kieran and Wagner 1989, p. 8), a goal that, in my opinion, has not been achieved.

To many authors, CAS is an appropriate technology for this "new balance" because it is not limited by the approximate treatment of numbers or by the necessity of programming. Mayes (1997) states that authors of research papers on CAS often study how CAS may help "set a new balance between skills and understanding" or "resequence skills and understanding."

In France, the picture is somewhat different because the French curriculum is centralized, forcing researchers to focus on integrating CAS into this curriculum rather than assess improvements of students' conceptual reflection in small-scale projects. For this reason, we directed our attention toward the changes produced by the introduction of CAS in teaching and learning mathematics in everyday situations and toward the search for conditions that produce satisfactory results.

Techniques in the CAS Context

Hirliman (1996) and Artigue (1997) documented our studies of the use of Derive in French classrooms from 1994 to 1996. In these studies, we interviewed teachers using Derive and asked their opinions about the support that CAS brings to teaching and learning mathematics. In the Derive experiment, teachers commonly assumed that CAS "lighten" the technical work of doing mathematics, thus focusing students' attention on application or understanding. In reality, our survey of the French classrooms in the Derive experiment showed neither a clear decrease in the technical aspects of the work nor a definite enhancement of students' conceptual reflection (Lagrange 1996). Technical difficulties in the use of CAS replaced the usual difficulties that students encountered in paper-and-pencil calculations. Easier calculation did not automatically enhance students' reflection and understanding. Furthermore, many students who liked problem solving using computer algebra did not perceive that CAS convincingly supported their understanding of mathematics.

From the foregoing observations, we began to think of techniques as a link between tasks and conceptual reflection rather than something we should eliminate from mathematics education. In this chapter, the term *technique* is defined in its broadest sense as a particular way of doing a particular task. A technique is generally a mixture of routine and reflection. It plays a pragmatic role when the important thing is to complete the task or when the task is a routine part of another task. Technique plays an epistemic role by contributing to an understanding of the objects that it handles, particularly during its elaboration. It also serves as an object for a conceptual reflection when compared with other techniques and when discussed with regard to consistency. Without CAS, paper-and-pencil techniques cannot be avoided because of their pragmatic role in mathematics education. Therefore, teachers and researchers tend to consider only their practical role, neglecting the epistemic contribution.

The teachers we observed in the French Derive experiment encouraged students to "jump" directly from tasks to a conceptual reflection; these teachers were unable to see that techniques specific to the use of symbolic computation are necessary and useful, that these techniques are not obvious to students, and that these techniques may be a topic for reflection. Because many students' understanding of mathematics was based on paper-and-pencil techniques used in the ordinary context, students believed that solving problems with CAS was not closely aligned with these techniques. Therefore, they did not consider Derive a useful tool in developing their mathematical understanding.

To explain a "technique use" of CAS and its pragmatic and epistemic role, I have chosen a problem developed by Mounier and Aldon (1996) related to the factorization of $x^n - 1$. They regretted that tenth-grade students, whom they asked to find general factorizations of the preceding polynomial, were restricted in their investigation by the difficulties of the calculations and notations associated with using paper and pencil. Mounier and Aldon (1996) tried to use Derive to liberate students "from the technical aspects of computing by hand" while encouraging them "to keep sight of the main goal." Observing a set of outputs from Derive's Factor Rational command, several students suggested that a general factorization should be

$$(x-1)(x^{n-1}+x^{n-2}+\ldots+x+1),$$

but other students objected, saying that this factorization is true only for odd values of n. These students had observed Derive's factorization for $n = 4$ and $n = 6$; they said that, with n even, the factorization should be

$$(x-1)(x+1)(x^{n-2}+x^{n-4}+\ldots+x^2+1).$$

Other students found that the suggested general factorization is not true for every odd n, because

$$x^9-1=(x-1)(x^2+x+1)(x^6+x^3+1).$$

Then all members of the class agreed that they should search for other general factorizations, and they did not find them. Although they did discover techniques for factorizations when n is prime or in the instance of a product of two primes, the theory of cyclotomic polynomials ensuing from these techniques is beyond the reach of tenth-grade students.

In my analysis, students found it conceptually difficult to grasp the following concept: for a given n, several factorizations may exist, but only one is "general" or true for every integer n. Using CAS, students encountered the difficulty explained above because Derive's Factor Rational command returns the most factored form of a polynomial and this form is "general" only for n prime. This situation is interesting because students have obviously learned much about algebra if they can say, "Yes, for given polynomials Derive gave us factored forms that are not the general factorization, but the general factorization is still true because we can find it by expanding parts in the factored forms." An important point to emphasize is that students who are able to make the statement in the preceding sentence obviously know what it means to select and expand a part in these forms. Mathematicians may think that this "technique" is obvious because they recognize complete and incomplete factorizations and understand that CAS provide the means to pass from one form to another. However, students were not easily able to select a part in an expression and expand it. In this situation, students' inability to manage factors in Derive could not be separated from their lack of understanding of the concept of factorization.

We observed that teachers in the Derive experiment were often reluctant to give time to techniques of managing expressions in Derive. Since working regularly in a computer room presented difficulties, most of the teachers taught only a few sessions with Derive. In this context, they saw little pragmatic use for Derive's techniques and tried—often unsuccessfully—

to focus on conceptual issues. In contrast, experienced teachers and researchers successfully integrated Derive's techniques into the average classroom. For instance, Mounier's and Aldon's students had access to Derive on laptops in the classroom and for personal work. These teachers assigned the factorization of $x^n - 1$ as "a long term problem." According to Mounier and Aldon (1996), students did learn Derive as a new tool *and* changed their image of the concept of factorization. These teachers recognized the need for building techniques using Derive and the epistemic role of these techniques in the understanding of algebra.

Considering the roles of techniques, such positions as "techniques are no more relevant in the days of CAS" or "now mathematics is just application" cannot be satisfactory. Of course in the days of CAS, routine use of techniques and drill could become less necessary, and reflecting on a technique could become more important than just practicing. We need to think more of techniques and understanding as an interrelation rather than oppose them. In the next paragraph, I present a general framework for understanding this interrelation.

A Didactic Approach to the Role of Techniques

Considering the evidence of students' difficulties and the emphasis put on problem solving and application, researchers in France were concerned that algebra might disappear from school mathematics because many people viewed the manipulation of symbols as meaningless. With calculators to numerically solve problems, why should teachers continue to ask students for algebraic resolution, a task that many students tend to fail? Researchers, including Mercier (1996), first emphasized the power of algebra and its potential place as a foundation of school mathematics and then reflected on the reasons for this poor position of algebra in the reality of students and students' mathematical activity. They emphasized that because of successive reforms, school algebra consisted of "scattered techniques." No tasks would be explored that these techniques could help to achieve, because less emphasis was put on proofs. Thus no reflection on these techniques would occur that could make them visible mathematical entities.

From the observations outlined above, researchers emphasized that a mathematical topic should exist in school as a consistent set of tasks, techniques, and conceptual reflection. Chevallard (1999) refers to these tasks, techniques, and conceptual reflections by the term *praxeologies* because they involve *praxis* and *logos*. Students first see the tasks as problems, progressively learning to solve them as they become more skilled, thus acquiring techniques in dealing with a topic. On the one hand, these techniques exist for the pragmatic resolution of tasks. On the other hand, their elaboration often implies epistemic reflection. Furthermore, in teaching and learning situations, students and the teachers are not interested in simply acquiring and applying a set of techniques. They want to talk about them and, therefore, develop a specific language that they can use to question the consistency and the limits of the techniques. In this way, both students and teachers reach the conceptual or theoretical level in the topic, giving the techniques their plain epistemic dimension. An appropriate use of CAS should not focus on eliminating the learning of techniques. It should, instead, take advantage of the availability of new techniques to foster students' conceptual reflection.

The TI-92 Experiment

After conducting the Derive experiment, we began a new experiment to assess the relevance of a symbolic calculator—the TI-92—for students' learning of precalculus in the science "stream" of French schools. We did not conduct a survey of classrooms practices, as we had done in the Derive experiment. Working with teachers in two classes during a two-year period, we built a pilot precalculus project and experimented with this project.

In the classes of teachers participating in the TI-92 experiment, students had access to the TI-92 calculators for the entire year. Both teacher and students had the choice of using or not using the calculator for lessons and for personal work, thus meeting the first prerequisite for productive use of the calculator. Our work was then to look at the joint development of techniques and understanding to determine what classroom situations could make efficient use of CAS in the classrooms described above.

We worked with the nationally established curriculum in algebra and precalculus. In the eleventh grade, this curriculum focuses on an approach to the notions of function, limit, and derivative. The curriculum emphasizes solving, proving, and explaining. The experimental classes were in ordinary schools and were expected to reach achievement levels similar to other classes. In the first year, the teachers submitted the lessons and we researchers observed classroom situations and the students.

In previous publications (Lagrange 1999a, 1999b), I described, in more detail, the design of the TI-92 experiment and the first-year observations. The evidence of successes and difficulties helped us design sessions for teachers who participated in the second year of the experiment. From results of the Derive experiment, we learned that students should have time to build CAS techniques for successful integration to take place. We also learned that CAS techniques should coexist with paper-and-pencil techniques in certain work and that classroom discussion adds a conceptual dimension to the techniques. To show the diversity of techniques and the role of CAS in students' learning, I subsequently discuss the results of the second-year experiment.

A Session on Basic Algebraic Techniques

Using a symbolic calculator to perform basic algebraic transformations is not always easy. An important factor is the notion of equivalence of expressions and of the need for awareness of the different equivalent forms of an expression. A TI-92 user meets automatic simplification when an expression is entered. The screen in **figure 15.1** displays an example of a puzzling phenomenon that occurs with an automatic simplification. Two obviously equivalent expressions are "simplified" in two radically different forms.

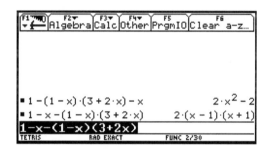

Fig. 15.1. A puzzling phenomenon that occurs with an automatic simplification

Students therefore cannot rely on automatic simplification for the form needed for an expression. In the isolated sessions of the Derive experiment, we often saw students searching for a given form of an expression by trial and error (Artigue 1997). This process was often unproductive and led to little mathematical insight. Because, in the TI-92 experiment, the goal was to use CAS on a regular basis and to foster mathematical understanding, we thought students needed to learn to consciously use the items of the algebra menu (Factor, Expand, ComDenom) to decide whether expressions are equivalent and to anticipate the output of a given transformation on a given expression. Thus, the goal of a lesson at the beginning of the year was to develop techniques in managing algebraic expressions using CAS. We developed a set of three tasks to help students acquire flexible use of the TI-92 commands of algebraic transformations for working on the equivalence of expressions.

The goal of the first task was for students to become aware of the output of the TI-92 and the many possible equivalent forms of an expression. In addition, we wanted students to remember the associated language. Students had to enter expressions and observe the TI-92 simplification.

They then had to identify the mathematical treatment that the simplification did. We chose the expressions (see **fig. 15.2**, left side of screen) to obtain a variety of simplifications (see **fig. 15.2**, right side of screen): expanding, factoring, reordering, partial fractional expanding, and cancellation by a possibly null expression.

```
┌F1────┐┌───┐┌───┐┌───┐┌─F5────┐┌────────┐
│▼ ƒ   ││Algebra││Calc││Other││PrgmIO││Clear a-z...│
└──────┘└───┘└───┘└───┘└───────┘└────────┘
 ■ (x - 1)·(x - 3) + x - 3              x² - 3·x
 ■ x - 3 + (x - 1)·(x - 3)              x·(x - 3)
     2   1                               1     2
 ■  ─ + ───                            ───── + ─
     x  x - 2                          x - 2   x
     2    x                              2    2
 ■  ─ + ───                            ───── + ─ + 1
     x  x - 2                          x - 2   x
       x - 2                                 1
 ■  ─────────                                ─
     x·x - 2·x                                x
 (x-2)/(x*x-2*x)
 MAIN         RAD EXACT         FUNC 10/30
```

Fig. 15.2. Student-entered expressions and their simplifications

In the second task, our goal was to link the understanding of general forms of expressions with the various items of the Algebra menu. Therefore, students explored the effect of various items of the Algebra menu on the expressions in **figure 15.2**. The number of transformed expressions depends on the expression. For instance, every transformation of

$$\frac{x-2}{x \times x - 2x}$$

gives

$$\frac{1}{x},$$

whereas Expand, ComDenom and Factor have different effects on

$$\frac{2}{x} + \frac{x}{x-2}.$$

Interesting discussion may follow this observation. In this task, the students also learned how to copy an expression into the entry line. Therefore, they saved time and effort by not entering expressions several times.

After completing this task, students could learn to use these transformations to decide whether two given expressions are equivalent. With CAS, the technique is as follows: enter the two expressions separated with the equals sign, and simplify. CAS generally returns "true" for equivalent expressions (see **fig. 15.3**.)

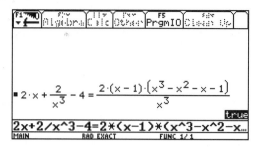

Fig. 15.3. A CAS technique to decide whether two expressions are equivalent

The epistemic value of this technique is poor because it lends no insight into the reasons for the equivalence. Although the technique is simple, it requires entering two expressions and thus can be tedious for heavy expressions. We designed the third task to make the use of transformations more convenient than entering a test of equality. The task gave a rational expression G along with three other apparently equivalent rational expressions H, I, and J. Students were to decide which expression was equivalent to G. We expected students to enter G and to find a TI-92 command to obtain a form similar to H, I, or J rather than perform three tests of equivalence, because entering tests is more tedious. Several similar tasks were offered to students.

To encourage students to use various transformations, we offered expressions in different forms: reduced, more or less developed, and factored. For instance, G was

$$\frac{x^2 - 6x + 2}{2x - 1},$$

H was

$$\frac{-11x + 4}{4x - 2} + \frac{x}{2},$$

I was

$$\frac{3}{4(2x - 1)} - \frac{x}{2} + \frac{11}{4},$$

and J was

$$\frac{(x + \sqrt{7} - 3)(x - \sqrt{7} - 3)}{2x - 1}.$$

With these expressions, a good strategy is to expand G. It yields to an expression opposite to I. The user can then copy this expression in the entry line, delete the *x*/2 term, and apply the ComDenom command. G and H are thus proved equivalent. J is a factored form of H, but the mere Factor command does not transform H into J (fig. 15.4). Because J is a "radical" factorization of H, a special form of the command must be applied.

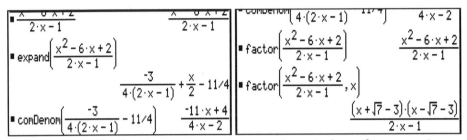

Fig. 15.4. Using the ComDenom command to prove the equivalence of two expressions

The resolution described is only one of several possible strategies. For instance, students might start from expression H or J. This topic can prompt many rich mathematical discussions in the classroom.

The students experimented with these tasks soon after they received their calculators. On the one hand, students did well with the tasks and the discussions. On the other hand, the work was not easy for eleventh-grade students and a one-hour session was too short. The second task was time-consuming, even when students copied teacher-made expressions into the entry line. The task had to be cut short to allow time for the third task. For the teacher, the challenge was thus to manage the time and to establish a balance between handling expressions in the calculator and conducting a mathematically productive discussion. In this session, the students were introduced to techniques that they could use in their everyday practice with the calculator. The hope was that these techniques might help the students better understand such notions as the equivalence of expressions or the forms of an expression, for instance, expansion and factorization.

Techniques for Generalized Problems

In France and other countries, problems of optimization are popular because students can work on optimization problems using precalculus concepts. The graphic and numerical facilities of calculators are excellent supports to encourage students to consider multiple approaches to these problems. In France, students learn derivatives in eleventh grade and can then solve optimization problems symbolically. They have learned various techniques to tackle these problems and are able to make sense of precalculus concepts when elaborating these techniques and reflecting on them.

Learning Techniques and Concepts Using CAS

By designing lessons for students using a TI-92, we thought that the availability of CAS could help amplify the tasks of optimizing and the techniques to pass from a particular to a generalized configuration. Consider the following problem:

> A man wants to build a tank. The walls and base of the tank are to be made of concrete 20 cm thick, the base is to be a square, and the tank must contain 32 cubic meters. Let x be the horizontal dimension of the side of the inner square, and let h be the inner vertical of the tank, both measured in meters. What should be the values of x and h to use as little concrete as possible?

The function giving the quantity of concrete simplifies into

$$\frac{25x^4 + 20x^3 + 4x^2 + 3200x + 640}{125x^2}.$$

Generally, students do a graphical or numerical study of this function and observe that a minimum seems to appear near the value $x = 4$ (see **fig. 15.5**).

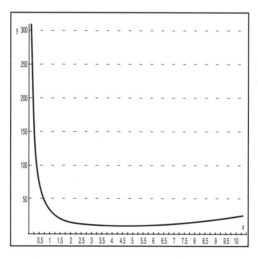

Fig. 15. 5. Students search for a minimum value to solve the concrete-tank problem.

In a more mathematical approach to the problem, students can compute, with or without the TI-92, the derivative of this function. They can find that it has two zeros, one for $x = 4$ and another for $x = -0.4$, and generally, they do a numerical study of the sign of this derivative, finding that it is negative between 0 and 4 and positive above 4.

This study is interesting, and the graphical, numerical, and even symbolic capabilities of the TI-92 can help. In my opinion, this approach has two significant limitations:

1. It primarily "encourages" the numerical or graphical approach to the problem. Students are not encouraged to use more powerful approaches, such as factoring the derivative to study its sign.

2. The answer is not interesting because it does not pave the way for new questions, such as "Is it a general result that there is a minimum and only one? What can you say of this minimum?" and so on.

Limitations such as those described above do not exist in the following generalized problem:

> The walls and base of the tank are to be made of concrete of the same thickness, e; the base is to be a square; and the tank's inner volume must be V. Let x be the horizontal dimension of the side of the inner square, and let h be the inner vertical of the tank, both measured in meters.

The aim of the problem is to know whether a value of x and h exists that uses as little concrete as possible. In addition, we want to know how this value depends on e and V. We first consider a tank with $e = 0.2$ meter and $V = 32$ cubic meters and look for a value of x and h to use as little concrete as possible. We then solve the general problem and answer the questions of independence.

With this generalized problem, the students could first understand the problem and try a numerical or graphical approach of the solution. They then considered the parametric function giving the quantity of concrete and its derivative. Many students found the task difficult because they could not conduct a graphical study of the function nor do a numerical study of the derivative like they usually did. They were driven to factor the derivative symbolically to study its sign. With the help of the TI-92, they obtained

$$\frac{2e\left((x+2e)\left(x-(2V)^{1/3}\right)\left(x^2+(2V)^{1/3}x+(2V)^{2/3}\right)\right)}{x^3}$$

and reflected on this expression to obtain that it has the same sign as

$$\left(x-(2V)^{1/3}\right)$$

for positive x and therefore that a minimum quantity of concrete is used at $x = (2V)^{1/3}$. They could answer yes, it depends on V, but no, it does not depend on e. This issue of dependence and independence is more important than the value of x itself.

This generalization, with the help of the TI-92, has the following advantages:

- The TI-92 capabilities are fully used, not only the graphing and numerical capabilities but also the symbolic calculations.
- The numerical problem helps students build a technique, fostering insight into optimization. This technique is not far from paper-and-pencil or graphical calculator techniques. The general problem is not simply a continuation of the numerical problem; it reveals the limitations of the existing technique and promotes a new, more general and symbolic technique.
- The objects that the general technique handles bring more sense to the problem.

The problem is an example of how CAS can provide new techniques in interrelation with paper-and-pencil techniques, opening a new understanding of optimization. These new techniques are possible because CAS allow students to interpret calculations with symbolic constants, or parameters, as a continuation of the same calculation with explicit values. As another example, Drijvers and van Herwaarden (2000) offer a report on students' solutions of two-variable systems and a discussion of the role of parameters and of techniques, or schemas, in the CAS context.

Conclusion

When calculators like the TI-92 became available, Waits and his colleagues (1997), among many authors, emphasized that today "we have better tools than paper-and-pencil exercises for doing the procedures of mathematics" and suggested that researchers work on "how to teach the next level above doing or practicing procedures—thinking about mathematics." In this chapter, I illustrated how thinking about a topic in mathematics could not be separated from tasks and techniques in exploring this topic. Tasks previously requiring time-consuming elaboration and tedious application through paper-and-pencil techniques can now be completed in a single keystroke. The potential of CAS in teaching and learning is not easy to appraise. On the one hand, teachers often assume that CAS will be efficient for more conceptual approaches. On the other hand, they are reluctant to give up paper-and-pencil techniques or graphical techniques. They believe that the epistemic role of these techniques is important for students to make sense of the underlying concepts, even when the pragmatic role of such techniques is challenged by CAS. Kendal and Stacey (1999) give the example of a teacher who privileged by-hand algebraic calculation in fostering students' understanding of notions in calculus and of another teacher who relied on graphical approaches. They show that integrating CAS is not easy for these teachers because using CAS could inhibit students' work on paper-and-pencil techniques or graphical techniques that the teachers consider important.

Through the examples in this chapter, I focused on techniques of using CAS and on students' work, showing their pragmatic importance as well as their possible epistemic contribution. Consequently, CAS should be considered as an opportunity to take advantage of the availability of these new techniques rather than a way to lessen the technical dimension in teaching and learning mathematics. This approach to the educational use of CAS opens new perspectives. The pragmatic and epistemic roles of paper-and-pencil techniques and CAS techniques help teachers in the delicate design of new classroom situations.

Situations like the "generalized tank problem" take advantage of CAS potentialities, especially calculations on generalized expressions and the possibility of rethinking a calculation by replacing numerical constants by symbolic parameters. On the one hand, these situations are relatively easy to integrate into a curriculum because they take advantage of the more obvious potentialities of CAS and because their goal is clear mathematical knowledge. On the other hand, they cannot be successful without the teacher's awareness of specific techniques and strong teacher involvement to guide students to understand the differences between numerical and "generalized" techniques.

Situations with a goal of using CAS efficiently, such as the example of a session on algebraic techniques, are more difficult to integrate into a curriculum because they seem directed toward the instrument rather than toward mathematics. However, not teaching these techniques increases obstacles in students' use of CAS. Their epistemic contribution is also important, provided that the teacher promotes the more meaningful techniques. The actual difficulty is to conceive the mathematical knowledge underlying the use of CAS in an educational context in which mathematical thinking remains primarily ruled by paper-and-pencil practices.

REFERENCES

Artigue, Michèle. "Le logiciel Derive comme révélateur de phénomènes didactiques liés à l'utilisation d'environnements informatiques pour l'apprentissage." *Educational Studies in Mathematics* 33(2) (1997): 133–69.

Chevallard, Yves. "L'analyse des pratiques enseignantes en théorie anthropologique du didactique." *Recherches en didactique des mathématiques* 19(2) (1999): 221–66.

Drijvers, Paul, and Onno van Herwaarden. "Instrumentation of ICT-Tools: The Case of Algebra in a Computer Algebra Environment." *International Journal of Computer Algebra in Mathematics Education* 7(4) (2000): 255–75.

Hirliman, Anne. "Computer Algebra Systems in French Secondary Schools." *International Derive Journal* 3(3) (1996): 1–4.

Kendal, Margaret, and Kaye Stacey. "Varieties of Teacher Privileging for Teaching Calculus with Computer Algebra Systems." *International Journal of Computer Algebra in Mathematics Education* 6(4) (1999): 233–47.

Kieran, Carolyn, and Sigrid Wagner. "The Research Agenda Conference on Algebra: Background and Issues." In *Research Issues in the Learning and Teaching of Algebra*, edited by Sigrid Wagner and Carolyn Kieran, pp. 1–10. Reston, Va.: Lawrence Erlbaum Associates and National Council of Teachers of Mathematics, 1989.

Lagrange, Jean-Baptiste. "Analysing Actual Use of a Computer Algebra System in the Teaching and Learning of Mathematics." *International Derive Journal* 3(3) (1996): 91–108.

———. "Techniques and Concepts in Pre-calculus Using CAS: A Two Year Classroom Experiment with the TI-92." *International Journal for Computer Algebra in Mathematics Education* 6(2) (1999a): 143–65.

———. "Complex Calculators in the Classroom: Theoretical and Practical Reflections on Teaching Pre-calculus." *International Journal of Computers for Mathematical Learning* 4(1) (1999b): 51–81.

Mayes, Robert. "Current State of Research into CAS in Mathematics Education." In *The State of Computer Algebra in Mathematics Education*, edited by J. Berry and J. Monaghan, pp. 171–89. Bromley, England: Chartwell-Bratt, 1997.

Mercier, Alain. "L'algébrique, une dimension fondatrice des pratiques mathématiques scolaires." In *Proceedings of the 8th summer school of didactique des mathématiques*, edited by Perrin-Glorian Noirfalise, pp. 345–61. Clermont Ferrand, France: IREM 1996.

Mounier, Georges, and Gilles Aldon. "A Problem Story: Factorizations of $x^n - 1$." *International Derive Journal* 3(3) (1996): 51–61.

Rachlin, Sidney L. "The Research Agenda in Algebra: A Curriculum Development Perspective." In *Research Issues in the Learning and Teaching of Algebra*, edited by Sigrid Wagner and Carolyn Kieran, pp. 257–65 Reston, Va.: Lawrence Erlbaum Associates and National Council of Teachers of Mathematics, 1989.

Waits, Bert, Franklin Demana, and Bernhard Kutzler. "Guest Editorial." *International Journal of Computer Algebra in Mathematics Education* 4(1) (1997): 4–6.

Activity 9

1. Find two numbers such that the sum of each and its reciprocal is $\frac{17}{5}$. Show that your answers are correct.

Solution:

Let x = a number

- solve$\left(x+\dfrac{1}{x}=17/5, x\right)$

$$x = \dfrac{3 \cdot \sqrt{21}+17}{10} \text{ or } x = \dfrac{^-3 \cdot \sqrt{21}-17}{10}$$

- $x+\dfrac{1}{x} \mid x = \left\{\dfrac{3 \cdot \sqrt{21}+17}{10} \quad \dfrac{^-(3 \cdot \sqrt{21}-17)}{10}\right\}$

$$\{17/5 \quad 17/5\}$$

2. A collection of 25 coins is worth $2.10. If the coins are nickels and dimes, how many of each coin are in the collection?

Solution:

Let d = the number of dimes, and let $(25 - d)$ = the number of nickels.

- solve($10 \cdot d + 5 \cdot (25-d) = 210, d$) $d=17$

- $25-d \mid d=17$ 8

- $10 \cdot d + 5 \cdot n \mid d=17$ and $n=8$ 210

The collection contains 17 dimes and 8 nickels.

Note: Writing the equation is the difficult part of solving story problems; therefore translating story problems into equation form remains an important skill that will not be taken over by CAS.

PART IV

CAS and Assessment of Mathematical Understanding and Skill: Introduction

AMONG THE most important practical dilemmas raised by the introduction of calculators and computers into mathematics classrooms are those related to assessing students' learning. For what types of tasks will students' responses give useful information about the mathematical skills and understandings that students have developed? What response formats will be both informative and practical to use? How should traditional assessment practices be adapted for use in classrooms in which powerful mathematics technology is routinely available? These questions are addressed in the three chapters of this section.

Lynda Ball and Kaye Stacey write about their experiences in Australia, where they have worked to find ways to encourage students to record the reasoning used in applying technology to mathematical tasks and to assess that reasoning. Raymond Cannon and Bernard Madison describe and analyze the experiences of including technology in assessments administered in the American Advanced Placement Calculus program. Finally, Lin McMullin addresses the practical classroom problems of adapting traditional conceptions and techniques of mathematics testing to the new realities of computer- and calculator-rich classrooms.

—*James T. Fey*

CHAPTER 16

What Should Students Record When Solving Problems with CAS?

Reasons, Information, the Plan, and Some Answers

Lynda Ball
Kaye Stacey

SCHOOL mathematics, as it is taught and tested, is a social practice of a community of teachers, students, and assessors, built on the basis of more fundamental aspects of the discipline of mathematics. This practice evolves in response to both social and technological changes. The widespread availability of calculators and computers equipped with a computer algebra system (CAS) is currently the driving force behind many changes that will result in slow adjustment of the practices of mathematics. The task for mathematics educators is to guide the direction of change. This chapter discusses one attempt to influence this change.

The chapter is prompted by a practical concern at the heart of daily mathematics teaching. When students are using a CAS, how should they record their solutions to the exercises and more substantial problems that they undertake in the course of learning, doing, and being assessed in mathematics? We must address this urgent issue because we are now experimenting with students who use CAS calculators as routine technology in the final years of school (Stacey, Asp, and McCrae 2000). In this chapter we describe our response to this concern. On the practical level, we have created a rubric to guide students, teachers, and assessors. We hope that this rubric will alleviate day-to-day concerns and encourage students to give more attention to mathematical reasoning. Behind the rubric explained in this chapter lies thinking that relates to the big issues underlying CAS use for school mathematics. What are sensible examples and problems to set?

What are good practices to encourage in schools? What does it mean to understand mathematics and to demonstrate that understanding?

The Purposes of Written Solutions

The first step in deciding what to ask students to record is to consider what a written record of a solution should achieve. The first purpose of a written record is to extend the capacity of short-term memory. As students work through mathematical problems, they may need to jot down information to use later in the solution—paper is a wonderful "technology" for achieving this goal. The second purpose of a written record is to communicate how a problem has been solved. The written record should give another person, or the same writer at a later time, enough information to replicate the solution to verify its correctness, learn how to solve similar problems, or use the record for assessment purposes. Paper is also an ideal technology for this purpose, but it is not well used for this goal. In our experience, students more often use the written record to help them remember information they will need later. Students tend to write down the steps they need to remember and focus on the details, but they devote less of their attention to written communication of their mathematical reasoning, the plan of attack, and the solution path.

Introducing CAS into the school mathematics environment upsets the current equilibrium and presents the opportunity for new practices to be established. Our goal is to shift the focus of students' written records from principally providing the detail to principally providing the overview and emphasizing the reasoning. Using the "fitting a cubic" problem as a starting point (see **fig. 16.1**), we have shown an example of a student's by-hand solution.

"*Fitting a cubic*" *problem:* A cubic, $f(x)$, passes through the x-axis at the point $x = 2$ and has a local minimum at $(-3, 0)$. Given that $f'(-2) = 3$, find $f(x)$.

In solution 1 (see **fig. 16.1**), all the significant reasoning is implicit and only the routine reasoning is explicit. We often see this type of solution in students' work. The first line, for example, does not explain the thinking that links the local minimum at $(-3, 0)$ to the repeated factor of $(x + 3)$. Readers must fill in the reasons for themselves. Yet the routine use of the product rule is explicitly noted, and the routine step of taking out the common factor of $a(x + 3)$ to obtain $f'(x)$ in factorized form is explicitly recorded. The written steps in solution 1 are instances of recording to extend short-term memory. Written student work similar to the solution in **figure 16.1** serves to extend short-term memory but does not communicate the mathematical thinking very well. This type of work is useful only to a

> **Solution 1**
>
> $f(2) = 0, f(-3) = 0,$ and $f'(-3) = 0 \Rightarrow f(x) = a(x - 2)(x + 3)^2$
>
> $f'(x) = a(x + 3)^2 + 2a(x - 2)(x + 3)$ using the product rule
> $= a(x + 3)(x + 3 + 2(x - 2))$
> $= a(x + 3)(3x - 1)$
>
> $f'(-2) = a(1)(-7)$
> $= -7a$
>
> $f'(-2) = 3 \Rightarrow -7a = 3$
>
> $a = -\dfrac{3}{7}$
>
> $f(x) = -\dfrac{3}{7}(x - 2)(x + 3)^2$

Fig. 16.1. A student's correct by-hand solution to the "fitting a cubic" problem

teacher or other knowledgeable reader who does not actually need an explanation of the thinking.

The introduction of CAS requires rethinking how students record their results, because much of what students currently consider as important in the written record will disappear. The calculation of the derivative of $f(x)$ in factorized form requires only one or two steps on current entry-level CAS calculators, such as the Texas Instruments TI 89, Casio FX 2.0, and Hewlett Packard HP 40G. Solving the equation $f'(-2) = 3$ is also a one-step operation. With more sophisticated technology, even fewer routine steps are required. When using Mathematica, just two lines of input and one line of output solve the "fitting a cubic" problem, as shown in solution 2 (see **fig. 16.2**).

The students and teachers with whom we work are usually perplexed about what they should write in the context when CAS is available. If routine steps of algebraic manipulation are seen as the most important feature of a written record of a solution, a dilemma now arises. With CAS these steps now occur within the machine and so may not be accessible. Even when the steps are accessible, transcribing intermediate detail from machine to paper is not good practice for using technology. Not having to write out all the routine steps presents new opportunities to communicate mathematical thinking. We hope to use this interruption to an established practice as an opportunity to intervene and to set new norms, so that students aim primarily to communicate their mathematical thinking rather than the technical details of a solution. Beyond solving a practical problem, this practice offers a number of benefits that we outline below.

RIPA: A Rubric for Writing Solutions

To assist students and teachers in making the change in the nature of written records, we have developed a set of guidelines for classroom use. We propose that good communication of mathematical thinking can be achieved if students—

- write down all their *reasons*;
- write down all the *information* that they use, including the input they entered into the CAS;
- make sure that the *plan* of the solution is clear so that the solution path can easily be followed; and
- write down selected *answers* only—not every intermediate answer need be recorded.

Thus, *reasons* (R), *information* and inputs (I), the *plan* for the solution path (P), and some of the *answers* (A) are the key ingredients—encapsulated by the mnemonic RIPA. These guidelines are summarized, along with brief notes of explanation and the implications for students, teachers, and assessors (see **table 16.1**).

Write down all reasons

We hope that using CAS frees students' attention so that they see past the details of calculations and manipulations and devote some attention to explicitly recording reasons. In this way, we hope to help bring mathematical connections and reasoning to the forefront. Currently, students primarily use their written record to extend short-term memory, using it principally for computational detail. Therefore, their reasoning is often implicit, as illustrated in solution 1 (see **fig. 16.1**).

However, encouraging students to write more reasons is not a straightforward task. Yackel (2001) and colleagues have observed and analyzed *explanation*, *justification*, and *argumentation* as social constructs in classrooms. Yackel (2001) points out that what counts as an acceptable explanation depends on the interactions of the students and teacher in the classroom. She describes how in a second-grade classroom, what counted as an explanation became more cryptic over time. Students began by giving "full" explanations, but over time more of the underlying reasoning became "taken as shared" and, hence, did not need explanation.

A similar phenomenon can also be observed in advanced mathematics classes. For example, typical student solutions such as solution 1 (see **fig. 16.1**) demonstrate cryptic and curtailed reasoning. As students practice applications of differentiation to curve sketching and work a wide variety of examples based on similar principles, most of the underlying reasoning becomes "taken as shared" with the teacher, assessor, or fellow students. As a result, when students write a solution, they see little need for recording their reasons. In reality, students write down few reasons to standard questions. The challenge for teachers and assessors is to pose questions that are accessible but sufficiently fresh so that students appreciate the need to specify the reasoning involved.

Make sure the plan is clear

Access to CAS increases the number of viable ways that students can solve problems, an observation noted often in related literature (National Council for Educational Technology 1994). Several different approaches to solving the "fitting a cubic" problem are available (see **fig. 16.1** and **16.2**). On the one hand, students working by hand have the advantage of carefully building the algebraic form of the function initially and thereby reducing the number of variables (i.e., beginning with $f(x) = a(x - 2)(x + 3)^2$). On the other hand, students working with CAS may find that less elegant methods can also be efficient. For example, solution 4 (see **fig. 16.2**), starting with $f(x) = ax^3 + bx^2 + cx + d$ and setting up four linear equations in the variables a, b, c, and d, is also a quick method. The knowledge that an increased range of viable solutions is possible with CAS may motivate students to communicate the plan of their solution clearly.

Because planning in mathematical problem solving does not occur in one step at the beginning of a solution but occurs opportunistically throughout in response to emerging findings (Hayes-Roth and Hayes-Roth 1979), we do not suggest that students write their plan at the start. Instead, they should look over their completed work and verify that the plan is clear.

Write down all information and inputs

Interestingly, the input into the CAS calculator appears to be more crucial to a good written record of the solution than the output. The functions, numbers, and equations that are entered need to be shown in the written record, as do all the operations, such as SOLVE, GRAPH, DIFFERENTIATE, and so on. Once again, this information might be recorded explicitly, for instance, "Substitute –2 into the derivative of $f(x)$," or implicitly, for example, "Find $f'(-2)$" or even "$f'(-2) = -7a$."

Teachers should encourage students to record their work using mathematical terminology rather than calculator-button sequences. In addition to producing better-written records, teachers can use this strategy to focus students' attention on the mathematics rather than on the technology. Although some discussion and writing in the classroom using calculator-

Table 16.1

Guidelines for Students, Teachers and Assessors on Record of Solutions (RIPA)

Reasons	Information	Plan	Answers	Benefits
Write down all reasons.	Write down all information and calculator inputs.	Make sure the plan is clear.	Write down selected answers.	
Notes for students				
• Give all reasons. • Carefully explain formulation of the problem into mathematical terms.	• Use mathematical notation rather than calculator syntax. • Illustrate information with graphs and diagrams. • Verify that reader should not need to know the question to understand your solution.	• Look over work to make sure the plan is clear. • Sometimes the plan is obvious from the reasons, information, and answers; sometimes it must be stated. • Give enough information so that someone else can use your method to solve a similar problem.	• It is not essential to write down all intermediate results—be selective. • Use mathematical notation rather than calculator syntax. • Write down your interpretation of the final results.	• Offers students guidelines for what is valued. • Promotes overview via the plan. • Minimizes transcription errors. • Facilitates checking.

Table 16.1—Continued

Reasons	Information	Plan	Answers	Benefits
	Notes for teachers to model good recording			
• Encourage students to record reasons. • Remember that the "level" of reasons depends on expertise and audience. • Emphasize short, written explanations.	• Use correct mathematical language and notation. • Teach links between calculator syntax and standard mathematical notation.	• Illustrate how plans can be shown both implicitly and explicitly. • Discuss examples with insufficient information or unnecessary information.	• Identify significant answers in examples, and explain why they are significant. • Value interpretation of results in context of questions.	• Provides framework to discuss written solutions. • Promotes emphasis on reasoning rather than the details of computation.
	Notes for assessors for question design and grading			
• Allocate partial credit for reasons with or without correct results. • Include questions that require formulation and novel reasoning.	• Expect correct mathematical notation. • Expect standard mathematical notation to evolve to include some calculator syntax.	• Expect and prepare for an expanded range of solving methods. • Set questions requiring the solution methods that are intended to be tested.	• Be explicit about answers that are required. • Don't expect intermediate results unless asked for. • Avoid artificial codes to indicate acceptable and unacceptable methods.	• Promotes student responses that are easy to grade because answers are stated explicitly. • Produces results justified with reasons. • Demonstrates logical mathematical sequence in plan.

button names may take place (e.g., press F8, then 3), this practice should be strongly discouraged in written mathematics. Inevitably, some technology syntax creeps into students' written mathematics, and teachers need to keep a "watching brief" on developments because syntax that is now unacceptable may become acceptable. Standard mathematical notation is likely to evolve by incorporating some technology syntax. Will such a command as DIFF($a(x - 2)(x + 3)^2$, x)|$x = -2$ for differentiating with respect to x and substituting $x = -2$ eventually become standard? As an interim guide, we propose that any calculator syntax that is recorded should apply across technology brands and not consist of keystrokes or notation associated with only one particular CAS.

Often an input into the calculator corresponds to several steps that would have been required using a by-hand method. The result is that CAS solutions are generally shorter than by-hand solutions. Solving a set of linear equations provides a good example of this phenomenon. We think students need record only the equations along with the fact that they are being solved, as is done in solution 4 of **figure 16.2**. Furthermore solution 2 of **figure 16.2** shows that it may not even be necessary to record the actual equations; it is sufficient to show where they come from. We are presently debating the necessity of recording steps that have been taken to "tweak" equations so that a particular CAS can solve the equation. For example, an equation may need to be rearranged before the CAS can solve it. Is recording such actions necessary? We are unsure. Technology constructs may also enter written mathematics in other ways. For example, recording a viewing window may become standard. Yet, for the moment, we believe that standard written mathematics should be preferred.

Write down only selected answers

During the process of a solution, several intermediate results are usually obtained before the final answer. In the "fitting a cubic" problem, intermediate results include the derivative $f'(x) = a(x + 3)(3x - 1)$ or another form, $f'(-2)$, the equation $-7a = 3$, and the value of

$$a = -\frac{3}{7}$$

before the final answer $f(x) = -\frac{3}{7}(x - 2)(x + 3)^2$ is obtained. In our opinion, if the reasons, the information, the input, and the plan of the solution are clear, students can choose which, if any, of these intermediate answers to record. In many short questions, students may not need to write down any output from the CAS calculator until they state their final answer. Our goal is to reduce "double handling" for situations in which students work with CAS but are constantly transferring information from machine to paper and possibly back again. Instead, teachers should encourage the good use of the "history" capabilities of the CAS. Just as good use of the memory is a hallmark of good scientific calculator use, good use of CAS is marked by good use of stored entries and outputs, thereby reducing re-entry.

The decision outlined above has an important consequence for assessors. If credit is to be given for an intermediate result, then students must be clearly instructed to provide it. Students may choose not to transcribe an unspecified intermediate result. In addition, when many solution paths are available for a given problem, an unspecified intermediate result may not arise, as illustrated in solutions 3, 4, and 5 (see **fig. 16.2**). This need for greater explicitness by assessors is now widely recognized in the literature on adapting examinations for CAS use (Rothery 1996).

It is not our intent to make rules about what students should not write down. Many situations will arise for which students will choose to record intermediate answers, and students should be allowed to make such choices. Technological change may also become a factor. In the future, integrated packages combining word processing and mathematical capabilities will be readily available to students. Mathematica, Scientific Notebook, and TI-Interactive are current examples that offer new ways for demonstrating solutions in multiple representations with written commentary.

Sample solutions discussed

Figure 16.2 shows five written records of solutions to the "fitting a cubic" problem. Solutions 3, 5, and 6 follow the same plan, whereas solution 4 is different. We think that solutions 3, 4, and 5 are good written records and that solution 6 is inadequate. In solution 3, the plan, although implicit, is quite clear, being apparent through the structure of the solution:

- Sketch the graph (omitted from **fig. 16.2**).
- Formulate a function incorporating as much given information as possible.
- Find the derivative.
- Determine a using $f'(-2) = -7a = 3$.

Solution 2

Define $f(x) = ax^3 + bx^2 + cx + d$

Solve $\{f(2) = 0, f(-3) = 0, f'(-3) = 0, f'(-2) = 3\}$ for a, b, c, and d

OUTPUT: $a = -\dfrac{3}{7}, b = -\dfrac{12}{7}, c = \dfrac{9}{7},$ and $d = \dfrac{54}{7}$

Solution 3

(Gives sketch graph first)

$f(2) = 0$, so $(x - 2)$ is a factor

$f(x)$ touches the x-axis at $x = 3$, so $(x - 3)$ is a repeated factor

Hence $f(x) = a(x - 2)(x + 3)^2$

$f'(x) = a(x + 3)(3x - 1)$

Substitute $x = -2$ into the derivative

$f'(-2) = -7a$

$f'(-2) = 3 \Rightarrow -7a = 3$

$a = -\dfrac{3}{7}$

$f(x) = -\dfrac{3}{7}(x - 2)(x + 3)^2$

Check by graphing

Solution 4

$f(x) = ax^3 + bx^2 + cx + d$ because a cubic

Use $f(2) = 0, f(-3) = 0, f'(-3) = 0,$ and $f'(-2) = 3$ to find 4 equations in 4 unknowns

$f(2) = 8a + 4b + 2c + d = 0$ [cuts x-axis at $x = 2$]

$f(-3) = -27a + 9b - 3c + d = 0$

$f'(x) = 3ax^2 + 2bx + c$

$f'(-3) = 27a - 6b + c = 0$ as min at $(-3, 0)$

$f'(-2) = 12a - 4b + c = 3$ as $f'(-2) = 3$

(Continued on next page)

(Continued from previous page)

Solving the system of linear equations using matrices: $\begin{bmatrix} a \\ b \\ c \\ d \end{bmatrix} = \begin{bmatrix} -\frac{3}{7} \\ -\frac{12}{7} \\ \frac{9}{7} \\ \frac{54}{7} \end{bmatrix}$

so $f(x) = -\frac{3}{7}x^3 - \frac{12}{7}x^2 + \frac{9}{7}x + \frac{54}{7}$

Solution 5

(Gives sketch graph first)

$f(2) = 0$, $f(-3) = 0$, and $f(-3)$ is a minimum gives

$f(x) = a(x - 2)(x + 3)^2$

Differentiate $f(x)$

Substitute $x = -2$ into $f'(x) = 3$ to determine a by solving the resulting equation

$f(x) = -\frac{3}{7}(x - 2)(x + 3)^2$

Solution 6

(Gives sketch graph first)

$f(x) = a(x - 2)(x + 3)^2$

$-7a = 3$

$a = -\frac{3}{7}$

$f(x) = -\frac{3}{7}(x - 2)(x + 3)^2$

Fig. 16.2. Five solutions to the "fitting a cubic" problem using CAS

- Write down the function substituting $a = -\frac{3}{7}$.
- Graph the function to check that each given property holds (not shown in **fig. 16.2**).

In solution 3 (see **fig. 16.2**) all the information from the question appears somewhere in the solution. The calculator inputs, which are an essential ingredient of explaining the plan, are the entry of $f(x) = a(x - 2) \cdot (x + 3)^2$ and differentiating and substituting to obtain $f'(-2) = 3$, both of which are recorded. The initial sketch graph, which is not shown, is also an important way of recording the information in the question. The most important reasoning in this solution is involved in establishing the form of $f(x)$, and this reasoning is adequately, although briefly, explained. Only a few intermediate answers are given. In comparison with the by-hand solution illustrated by solution 1 (see **fig. 16.1**), the intermediate forms of the derivative and the substitution of $x = -2$ were not recorded, but this omission does not detract from a reader's ability to follow the solution. Using a TI 89, FX 2.0, or HP 40G simplifies getting the derivative of $f(x) = a(x - 2)(x + 3)^2$ as $a(3x - 1)(x + 3)$. All three calculators differentiate the function with one button press, although obtaining the factorized form requires an additional button press on two of the calculators. In the future, students may not simplify expressions to the same degree as is currently expected. In this example, students gain no advantage by finding the derivative as $a(3x - 1)(x + 3)$ rather than $a(x + 3)^2 + a(x - 2)^2(x + 3)$, and teachers have no reason to insist on it. Solution 5 (see **fig. 16.2**) follows the same plan, but the description is more verbally explicit while providing fewer technical details. We believe that solution 5 is also an adequate written record.

Solution 4 (see **fig. 16.2**) follows a different plan. This plan is quite clear: use the information given to solve for the four unknown coefficients of the cubic. Solving systems of linear equations can now be done using one built-in feature of the calculator and should be considered a routine procedure. Therefore, we would not expect students to show any intermediate steps. If a method was used in which the inverse matrix was found and then used to find a solution, we still would not expect students to record the inverse matrix, because it is only part of an intermediate calculation. By using CAS, students can avoid many of the transcription errors that occured in the past when they used matrices.

Solution 6 (see **fig. 16.2**) illustrates what we believe is an inadequate written record. The work does not clearly illustrate how the results were obtained, and no reasons are given for any of the work. For example, no mathematical justification is given to explain why the student wrote $-7a = 3$. Students should be discouraged from giving such written records as solution 6, in which results are stated without any reasons. Only a reader who has already independently solved the problem can readily understand the solution method that the student used.

The use of CAS is advantageous for the "fitting a cubic" problem, but the question is certainly not trivialized by it. Students need to understand the relationship between a function and its graph and derivative to use CAS to answer the question. They should recognize that a local minimum on the x-axis means the existence of a repeated root and that $(x - 2)$ is a factor because the graph cuts the x-axis at $x = 2$. CAS can be used to differentiate and substitute values into the functions, but students need to identify the information in the problem and link this information to the structure of the expression for $f(x)$ to be able to solve the problem easily. Careful consideration of the structure of $f(x)$ with its repeated root demonstrates a good understanding of the mathematics associated with the problem. Solution 2 and solution 4 (see **fig. 16.2**) show that the "fitting a cubic" problem, as currently given, does not now require this understanding. Assessors need to be more explicit if this understanding is the outcome they want to test.

How to use the RIPA rubric

The RIPA rubric has been designed for class use to encourage students to produce written records of solutions in a way that highlights their mathematical thinking and is "robust" against advancing technology. By using students' own work and specially prepared solutions as illustrations, teachers can use the rubric to talk to students about how they record solutions. Teachers need to explicitly state what constitutes a good written record of a solution for a problem. In class, they can include examples that illustrate plans that do not contain enough information as well as those with unnecessary information. Students can be given a written record of a solution and be asked to produce a question that this solution answers. If the record does not have sufficient information, then students will not be able to complete this task or may need to recognize that multiple questions for the one solution might also exist. Teachers can then show the original question and hold a discussion about what can be added to the written record of the solution to adequately communicate the solution.

Using another approach, teachers can give students written records containing much more information than is necessary and ask them to identify the information that could be omitted. This instructional method highlights the need for all four features of the written record: the reasons, the information and inputs, a clear plan, and selected answers. The nature of questions asked in the classroom may change quite significantly as teachers produce tasks that enable students to develop this skill of recording solutions.

The rubric is also appropriate for use by assessors. Developing good questions to assess mathematical understanding using CAS requires new skills. As discussed elsewhere (McCrae and Flynn 2001), the availability of CAS trivializes some questions. More directly, the RIPA rubric points out that special care needs to be taken in allocating partial credit in constructed-response questions. Less variety is possible in solution paths and plans without technology, so more uniformity occurs in the intermediate answers that students obtain. In addition, these intermediate answers are recorded for future use. Yet in the CAS environment, the machine can carry out the short-term memory function, so the evidence for intermediate results may not exist. As a result, assessors need to be explicit about what intermediate results they want. A final potential benefit may be a more explicit recognition of the reasons that students give through the allocation of partial credit.

Conclusion

The purpose of the RIPA rubric is to assist students and teachers as they decide what to write in an environment in which some of them may think that nothing needs to be said. In accomplishing this goal, we hope to change the way students record their work to give more prominence to mathematical thinking. It is not our aim to restrict what students write. Using paper and pen in concert with modern technology is powerful, but the transcription of every intermediate answer is error-prone and should be avoided. It is not our aim to define hard-and-fast categories of reasons, inputs, and information, because these categories often overlap. Our message—a very important message—is to encourage students to include reasons, information, and inputs in their written records, not to carefully analyze their category. We have shown that the introduction of CAS interferes with existing school conventions and have suggested an intervention that might establish better practices.

REFERENCES

Flynn, Peter, and Barry McCrae. "Issues in Assessing the Impact of CAS on Mathematics Examinations." In *Numeracy and Beyond, Proceedings of the 24th Annual Conference of the Mathematics Education Research Group of Australasia*, edited by Janette Bobis, Bob Perry, and Michael Mitchelmore, pp. 210–17. Sydney: MERGA, 2001.

Hayes-Roth, Barbara, and Frederick Hayes-Roth. "A Cognitive Model of Planning." *Cognitive Science* 3 (1979): 275–310.

National Council for Educational Technology (NCET). *The A-level Curriculum of the Future—the Impact of Computer Algebra Systems*. Coventry: NCET, 1994.

Rothery, Andrew. "Using Derive in Calculus Exams." *Proceedings of the International Derive and TI-92 Conference*, edited by Baerbel Barzel, pp. 437–42. Schloß Birlinghoven: Zentrale Koordination Lehrerausbildung 1996.

Stacey, Kaye, Gary Asp, and Barry McCrae. "Goals for a CAS-active Senior Mathematics Curriculum." In *Proceedings of TIME 2000, an International Conference on Technology in Mathematics Education, December 11–14, 2000*, edited by Michael O. J. Thomas, pp. 244–52. Auckland, New Zealand: University of Auckland and Auckland University of Technology, 2000.

Yackel, Erna. "Explanation, Justification and Argumentation in Mathematics Classrooms." In *Proceedings of the 25th Conference of the International Group for the Psychology of Mathematics Education*, Vol. 1, edited by Marja van den Heuvel-Panhuizen, pp. 9–24. Utrecht, The Netherlands: PME, 2001.

Activity 10

Prove that any point on the perpendicular bisector of a segment is equidistant from the endpoints of the segment.

Solution:

Let the endpoints of the segment be $(0,0)$ and $(2a,0)$; then the line $x = a$ will be the perpendicular bisector of the segment. Any point on this line is represented by (a, y).

- $\sqrt{(a2-a1)^2 + (b2-b1)^2} \rightarrow \text{dist}(a1,b1,a2,b2)$ Done

- dist $(a,y,0,0)$ $\sqrt{y^2 + a^2}$

- dist $(a,y,2 \cdot a,0)$ $\sqrt{y^2 + a^2}$

The distance from (a, y) to $(0, 0)$ is the same as the distance from (a, y) to $(2a, 0)$. QED.

Chapter 17

Testing with Technology: Lessons Learned

Raymond J. Cannon
Bernard L. Madison

THE OBSERVATIONS, examples, and opinions presented in this chapter are based on our first-hand experiences with the Advanced Placement (AP) calculus examinations. Having served in the 1990s both as Chief Faculty Consultant—formerly Chief Reader—for AP calculus examinations and as members of the Advanced Placement Calculus Development Committee, we directed the scoring of all AP calculus examinations, set the cut scores for AP calculus grades, facilitated decisions regarding the AP calculus examinations and course descriptions, and made policy recommendations regarding the use of technology. As we see it, the relationship between AP calculus and the use of handheld calculators on AP calculus examinations has experienced four phases:

- years when scientific calculators were allowed on the examinations, 1983–1984;

- years when scientific calculators were required, 1993–1994;

- years when graphing calculators were required and a few graphing calculators with CAS were allowed, 1995–1998; and

- years when graphing calculators were required and several graphing calculators with CAS were allowed, 1999–present.

On the basis of our experience with calculators that possess graphic, numeric, and symbolic capabilities, we discuss all three uses and the interplay among them. Although the acronym CAS usually stands for *computer algebra system*—software that performs symbolic manipulation—we also use

CAS to represent a computer mathematics system that has graphic, numeric, and symbolic capabilities.

Although most of our experience with students' use of handheld calculators has been associated with assessment in calculus, many of the observations that we make in this chapter also apply to the use of calculators or computers in other courses, including algebra and trigonometry.

"With technology, some mathematics becomes more important, some mathematics becomes less important, and some mathematics becomes possible," summarized Henry Pollak, retired, of Bell Laboratories, quoted in *Crossroads in Mathematics: Standards for Introductory College Mathematics before Calculus* (1995). In this chapter we list ten "lessons learned" and include examples that illustrate the lessons—examples modeled primarily after AP calculus examinations and examples presented at the 1997 Technology Transitions Calculus Conference in Dallas, Texas (Madison, forthcoming).

To give context to our examples, we first present a short description of the AP calculus program, the role that calculators have played on the AP calculus examinations, and the scoring of those examinations. We then discuss the ten lessons we have learned and conclude with comments on attitudes of faculty members and the future.

AP Calculus and Calculators

Advanced Placement (AP) calculus is a program, sponsored by the College Board, that provides course descriptions and national examinations for two courses, calculus AB and calculus BC. Calculus AB is typically the first semester of college calculus. Calculus BC, an extension of calculus AB, is typically the first two semesters of college calculus. The College Board contracts with Educational Testing Service (ETS) to develop and score the AP calculus examinations.

During the period of time considered in this chapter, 1995–2000, the examinations for calculus AB and calculus BC consisted of two sections: forty-five multiple-choice items and six free-response questions. The two sections were equally weighted, and each was scored on the basis of 54 points. Thus, with 108 total points on the examination, each multiple-choice item had a value of 1.2 points and each free-response question had a value of 9 points.

The AP calculus free-response examinations are scored by college and school faculty who convene seven days each June during a session referred to as the *Reading*. During the Reading, a scoring standard is written for each free-response item, an outline of expected solutions is prepared, and a plan is developed to determine how to award points.

This chapter focuses on the approximately eight hundred fifty thousand examinations administered during the years 1995–2000. During those years, graphing calculators were required on some portions of the AP calculus examinations and not allowed on other portions of the examinations. Each year, a list of "approved" calculators was published; several models on the approved list had symbolic capabilities. Interestingly, in 1999 and 2000 the use of CAS calculators on the examinations increased dramatically, probably because the TI-89, Casio 9970, and other models were added to the approved list. Prior to 2000, only the multiple-choice sections of the examinations included a noncalculator portion, but the free-response section was divided into a calculator and a noncalculator portion in 2000.

The variety of student responses and other issues introduced by students using technology on the examinations has complicated the scoring process at the Reading. With more variety in responses, more variety is recognized on the scoring standards. *Readers*—the school and college faculty members who score examination papers—are made aware of the different approaches taken by students to avoid reading different responses as incorrect ones. For the readers, the most difficult and time-consuming part of the scoring process involves items that require students to justify or explain processes in which calculators are used to obtain results.

In the early 1980s, pressure was placed on The College Board and Educational Testing Service to allow calculators on their standardized tests. In response, the two groups allowed students to use scientific calculators on the AP calculus examinations in 1983 and 1984. This experimental phase lasted only two years because observers noted that students were not skilled in using the calculators; therefore, students were making mistakes on the examinations as a result of calculator use. Furthermore, scoring of the examinations was complicated because some students were using calculators and others were not. From the two-year experiment, educators involved with the examinations learned several lessons that later affected the calculator policies for AP calculus.

1. Calculators must either be required or be banned.
2. Calculators must play an essential role in students' learning calculus and completing the examination.
3. Students must be proficient in using calculators.

Less than a decade later, advances in calculator technology and the appearance of the reform movement in calculus created the opportunity for educators to apply the lessons outlined above.

In the mid 1980s, the movement to reform calculus instruction in colleges and universities had spread from small conferences, such as those held in 1983 at Williamstown College and in 1986 at Tulane University, to the large national Calculus for a New Century colloquium held in 1987, at the National Academy of Sciences as part of the project titled Mathematical Sciences in the Year 2000 (Steen 1998). In 1989, the National Council of Teachers of Mathematics (NCTM) published its *Curriculum and Evaluation Standards for School Mathematics* (NCTM 1989), which endorsed the use of calculators in teaching school mathematics. The calculators of choice in 1989 were "pure visualizers," the early programmable graphing calculators typified by the Texas Instruments TI-81.

In 1989–1990, against the backdrop described above, the AP Calculus Development Committee began deliberating its recommendations on the use of calculators. Issues of equity and college faculty attitudes were major considerations, as were pedagogical and assessment issues. Most of the discussants believed that calculus learning could be effectively assessed in almost any technology environment; they also believed that the environment must be equitable and that the results of the assessment must be credible and receive acceptance from college faculty who award credit for AP calculus work. Most people involved in the discussion agreed that the examination process could be effective with or without calculators, but if calculators were to become an essential part of the course, they must be an essential part of the examination. The discussants also believed that allowing students to learn with a tool that was not permitted in the examination was somewhat "improper" (Monaghan 2000).

In 1990, ETS conducted a major calculator-impact study (Morgan and Stevens 1991). Approximately seven thousand high school students from more than four hundred high schools participated in the survey. People associated with hundreds of colleges were also surveyed to determine their reactions to examinations that contain calculator-active questions, especially in regard to credit and placement. In view of this study, scientific calculators were required for the 1993 and 1994 AP calculus examinations, and the required use of graphing calculators was planned for the 1995 examinations. Another survey was conducted in 1993 to determine the feasibility of requiring graphing calculators in 1995, and tests containing graphing calculator items were administered to approximately three thousand students. The survey of colleges showed far fewer concerns about allowing graphing calculators than about allowing calculators with CAS. In view of all the survey and test results, graphing calculators were required on the 1995 AP calculus examinations. Some of the approved calculators had CAS capability, most notably the Hewlett Packard HP28 and HP48. Some calculators with CAS, such as the TI-92, were not allowed for various reasons; calculators with standard typing (QWERTY) keyboards were disallowed, partly for reasons of test security.

For the 1999 AP calculus examinations, other calculators with CAS had appeared and were added to the approved list, notably the TI-89 and Casio 9970. Several other CAS models have been added since that time.

The 1990 and 1993 surveys of colleges gave reason to believe that the use of CAS would weaken the support of AP calculus among the collegiate faculty. The surveys revealed more concern that symbolic manipulation would become mechanized than that graphing would become mechanized. Part of this difference in concern is caused by the way we have valued symbolic manipulation in assessing calculus knowledge. Being able to differentiate and antidifferentiate has been a major part of traditional calculus, as has been algebraic manipulation of derivatives and integrals. Since 1995, in an effort to form a hedge against the concerns of collegiate faculty and to prepare AP students for both noncalculator environments and calculator environments, the AP examinations have been divided into two parts, with one part requiring graphing calculators and the other part not allowing graphing calculators. From 1995 to 1999, only the multiple-choice section was so divided, but in 2000, when several new CAS calculator models were approved for use on the examinations, the free-response section was also divided. The decision to divide the examinations was not made to enhance assessment but to satisfy two groups with differing views regarding how students learn calculus best.

Writing good calculator-active test items was a major challenge for the AP calculus test development committee. The committee soon learned what others have learned (Monaghan 2000), that writing questions that bypass the technology is easier than writing questions that use the technology. In 1990, questions were written for the most advanced graphing devices available at that time, usually pure "visualizers" without solve keys or integral keys. Advances in technology made many of these questions no longer appropriate for the 1995 examination when graphing calculators were first required. We did recognize that the HP28 and HP48 had symbolic manipulation capabilities, but those features were somewhat discounted by the low student use of HP calculators and by the complexity of using these calculators. Students who used the HP machines made slightly better grades, but several other factors contributed to those higher grades.

Today, equity remains an issue because students use a variety of different calculators, some with CAS and some without. The functionality of the TI-89 is vastly different from that of the TI-81. We are relearning our lessons from 1983–1984! The practice of allowing, but not requiring, specific computer capability complicates assessment and makes inequities very difficult to avoid. Many people (e.g., Hurley [1999]) thought that the move from not allowing calculators to requiring them was a drastic move on the part of the AP program. However, this view is not substantiated. The move from prohibiting calculator use on the examinations to requiring calculator use is not as complicated as the option of allowing but not requiring CAS use on the examinations. This latter "in-between" option reflects the position currently taken by the AP calculus program.

Equity and acceptance, rather than assessment, have steered most decisions associated with the implementation of calculator use on the AP calculus examinations. Has the collegiate faculty accepted the use of CAS as a sound educational tool? Some have, and some have not. In response to the various levels of acceptance, the AP calculus committee has divided the examination into calculator and noncalculator parts.

Lessons Learned

Lesson 1
Good examination questions that make essential use of technology are difficult to write. Examination questions that neutralize technology are easier to write.

The first significant observation that test developers make after beginning to write examination questions for students who will use numeric, graphic, or symbolic technology is summarized above in lesson 1 (Monaghan 2000). Scientific calculators were allowed on AP calculus examinations in 1983–1984, but their only use, beyond performing arithmetic computations, was to change the form of answers. For example, $e^2 \sin 3$ could be rewritten as 1.04274. Such use of technology is peripheral to solving the problem, contributing only an added step after the problem is solved, but such practice does stimulate discussion regarding appropriate answers, as explored in lesson 10 of this chapter.

The following example requires the use of technology to solve the problem. A version of this example appeared on the AP calculus examinations soon after graphing calculators were required in 1995 and is reprinted below with permission of the College Entrance Examination Board.

Example 1.1
Find the area of the region bounded by the graphs of $y = x$, $y = \cos x$, and $x = 0$.

Typically, a student uses the calculator to draw the graphs, identify the region, and solve the equation $\cos x = x$. Then the area of the region is the integral $\int_0^A (\cos x - x)\, dx$ where A is the solution to $\cos x = x$. The step of finding a numeric approximation to A is precisely where technology is required. Students with some early models of graphing calculators, such as the TI-81, would need to find A by some method, for example, zooming and tracing; those students using more recent models would be able to use a solve key. Furthermore, similar differences apply to ways of evaluating the definite integral using a calculator. Some graphing calculators have no built-in method; some can evaluate it numerically, and those with CAS can evaluate it by antidifferentiating. In this example, the antiderivative is basic, so technology is not required; however, those who use their calculators do not have to risk making an error in remembering the antiderivatives. Often considered inequities, such issues are discussed later in this chapter. Questions regarding how accurate the approximation of A must be to achieve the requested accuracy of the final answer and what constitutes a final answer also follow. Other examples of items that make essential use of technology appear later in this chapter. A more difficult version of **example 1.1** is shown below as **example 2.1**.

Many ways of neutralizing various capabilities of computing technology exist and are important for two reasons: (1) removing or mitigating the advantage of students who have calculators with more powerful built-in features than those available to other students, and (2) testing students' abilities to do procedures that can be performed by technology. Most methods of neutralizing technology involve describing functions in ways other than giving an analytical formula. As examples, one can neutralize the power of CAS as follows:

- Give the graph of the derivative of a function, and ask for information about the behavior of the function. For example, ask for the location of the maxima and minima.

- Give a table of values for a function, and ask for information about the function.

  Note: In the absence of an analytical description of a function, students may attempt to use a polynomial fit to the given graph or data.

- Give students what would normally be the answer, that is, the result of taking a derivative or performing some algebraic expansion or simplification. Ask for a sequence of algebraic manipulations that lead to the result, or ask for something that uses the result, as illustrated in **examples 3.1** and **3.2**.

- Give properties of a function, such as that the function f is continuous and increasing. Ask for information about f.
- Describe a physical situation that can be modeled by an integral. Ask for the setup of the integral but not for its numerical value.
- Specify a method. For example, ask students to find an antiderivative by using partial fractions or to solve a differential equation by separation of variables. Of course, specifying a method to avoid CAS leads to another assessment question. If a student uses a different method, but not CAS, how will the work be evaluated? For example, what happens if the student uses an integrating factor to solve the differential equation?
- Introduce parameters. This technique is especially effective in neutralizing graphic-numeric technology.

Part of the reason that writing questions that bypass the technology is easier than writing questions that make use of the technology is that we have too little experience using the technology. Perhaps those educators who use this technology throughout their education will be better at understanding its power and determining where it can be used most effectively in the curriculum (Monaghan 2000).

Lesson 2

What you value in students' responses will change from computations and manipulations to analysis, setups, interpretations, and justifications.

Whether the use of CAS enhances mathematics education is actually a question relating to essential mathematical competences. Prior to the availability of efficient arithmetic calculators, most educators believed that the mastery of arithmetic computations was an essential competence. On the one hand, most high school or college-level instructors believed that arithmetic mistakes were minor mistakes—one-pointers. On the other hand, some instructors believed that mistakes were mistakes and incorrect answers received no credit. Algebraic mistakes were generally considered more serious, and differentiation mistakes were clearly the most serious. In the presence of a graphing calculator with CAS, algebraic manipulation and differentiation now tend to be on par with arithmetic computations. Consequently, competence in using the calculator with CAS can replace, to a large extent, competence in arithmetic and algebraic manipulation. What, then, is left to evaluate when assessing mathematical competence?

We next describe two examples that show how the introduction of calculators with CAS changes the way in which answers are assessed.

The following item appeared on the 2000 AP calculus examinations and is reprinted here with the permission of the College Entrance Examination Board. Because the following problem can be answered using a numerical solver and a numerical integrator, it "levels the playing field" for students with CAS and those without CAS.

Example 2.1

Let R be the region in the first quadrant enclosed by the graphs of $y = e^{-x^2}$ and $y = 1 - \cos x$ and by the y-axis, as shown in the figure here.

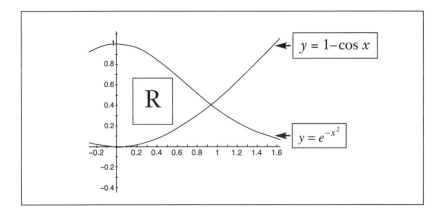

(a) Find the area of the region R.

(b) Find the volume of the solid generated when the region R is revolved about the x-axis.

(c) The region R is the base of a solid. For this solid, each cross section perpendicular to the x-axis is a square. Find the volume of this solid.

Graphing calculator technology is essential in solving the problem in example 2.1. First, the student must find the x-coordinate of the point of intersection of the two graphs. To do so, the student must solve $e^{-x^2} = 1 - \cos x$ and cannot accomplish this task by elementary algebraic methods. So the calculator solves it numerically. A value of 1 point, out of 9, is given to a student who finds the correct solution and correctly uses it as a limit on an integral in one of the three parts. Notice that the student must decide, in this instance, what the "correct" solution is. How accurate does this solution need to be to have three-place decimal accuracy in the three integrals to be evaluated? A more detailed discussion of the required accuracy on the AP examinations appears in lesson 10 of this chapter.

For part (a) of **example 2.1**, the student must evaluate

$$\int_0^A \left(e^{-x^2} - (1-\cos x)\right) dx$$

where A is the x-coordinate of the point of intersection of the graphs. This task can be done numerically, but not symbolically with elementary functions, so the student who is trying to use CAS encounters more difficulties than the student using a numerical integrator. Part (a) of **example 2.1** was valued at 2 points, 1 point for the integrand and 1 point for the answer. Only 1 point was at stake on the limits, and this point could be earned by having correct limits on any of the three integrals. The value of A, to six decimal places, is 0.941944. The value of the integral, to three decimal places, is 0.590 (truncated) or 0.591 (rounded). Using $A = 0.9$ gives the integral's value as 0.590 (rounded). In this example, the value of the integral is not very sensitive to the accuracy of A; however, if the graph of the integrand is steep near the limit of integration, then the value of the integral can easily be affected. How should we score

$$\int_0^{0.9} \left(e^{-x^2} - (1-\cos x)\right) dx = 0.590$$

as a solution to part (a) of **example 2.1**? The answer is as accurate as was requested, but the use of 0.9 as the upper limit is worrisome. To not give full credit, 2 points out of 2, the evaluator would be making the assumption that the student is careless and lucky, yet the student could be knowingly efficient.

For part (b) of **example 2.1**, the student must evaluate

$$\pi \int_0^A \left((e^{-x^2})^2 - (1-\cos x)^2\right) dx,$$

thus creating another grading dilemma. Part (b) was valued at 3 points, 2 points for the integrand and constant and 1 point for the answer. If three-place accuracy in answers is required, as on the AP examinations, how should we score the following response to part (b)?

$$\pi \int_0^{0.94} \left((e^{-x^2})^2 - (1-\cos x)^2\right) dx = 0.556\pi = 1.747$$

Is 0.555≠ a "correct" answer under AP rules? The dilemma here is that the 0.556 was obtained by rounding, but the 0.555 would be obtained by truncating the decimal expansion of the integral, and 0.555≠ is, in fact, not correct to three decimal places when the decimal approximation to ≠ is used. As a further irritation, keep in mind that

$$\pi \int_0^A \left((e^{-x^2})^2 - (1-\cos x)^2 \right) dx$$

is a number, and we are allowing as "correct" values of this number both 1.746 and 1.747. How would you respond if a student concluded that 1.746 = 1.747?

Another difficult approximation situation occurs if the student decides that to use CAS, an approximation of e^{-x^2} is needed using a Taylor polynomial, such as

$$1 - x^2 + \frac{x^4}{2} - \frac{x^6}{6} + \frac{x^8}{24}.$$

Does this student carry the burden of showing that the degree of the polynomial approximation used is sufficient for obtaining the required accuracy in the final answer?

For part (c) of **example 2.1**, the student must evaluate

$$\int_0^A \left(e^{-x^2} - (1-\cos x) \right)^2 dx;$$

no new issues are introduced here. Part (c) of **example 2.1** was valued at 3 points, 2 points for the integrand and 1 point for the answer. Notice that no points are allocated for algebra, and no points are allocated for differentiation or antidifferentiation.

The following problem is typical of a routine graphical analysis problem prior to the use of graphing calculators. For **example, 2.2**, we have assigned an arbitrary number of points so we can discuss how the presence of technology changes the grading process.

Example 2.2

Let f be the function defined by $f(x) = 2x^3 - 3x^2 - 36x + 5$.

(a) On what intervals is f increasing? Use the derivative of f to justify your answers. (3–4 points)

(b) Find the local extrema of f. Use derivatives of f to justify your answers. (4–5 points)

Prior to using graphing calculators, part (a) of example 2.2 would have been solved by taking the derivative,

$$f'(x) = 6x^2 - 6x - 36,$$

and such a process would have been valued at 1 or 2 points, out of 9 points.

After CAS, 1 point could have been given for recognizing that the derivative was positive, but no points would be given for actually computing the derivative. So the scoring standard might have read, "1 point for $f'(x) > 0$." This point could be awarded for either seeing this inequality explicitly set down or seeing it used implicitly in the student's work. Choosing the correct intervals, $(-\infty, -2]$ and $[3, \infty)$, would merit 1 or 2 points with or without CAS. Finally, for part (a), 1 point would hinge on noting that the derivative is positive on the intervals $(-\infty, -2)$ and $(3, \infty)$.

The student with a graphing calculator can graph the derivative, in numerical form if not using CAS and in symbolic form if using CAS, and can read the answer directly from the graph. Therefore, that student must translate into mathematical language the reason that this process gives the answer to the question. Such translation must be evaluated to score the student's answer. A discussion regarding the difficulties associated with assessing such work can be found in lessons 4, 5, and 6 of this chapter.

Previously, using only paper and pencil, a student would first find the zeros of f' and then decide where $f'(x)$ is positive and where it is negative. Doing so entails solving a quadratic equation and two quadratic inequalities. Such algebra might be valued at 2 or 3 points. The student would probably finish by justifying the answers using the first derivative test or second derivative test.

For part (b) of **example 2.2**, the student with the graphing calculator can graph the function and ask the calculator for the maximum (in an interval about –2) and the minimum (in an interval about 3), thereby obtaining the answers. The student is now under the obligation to give mathematical reasons why the chosen values are extrema, using either the first derivative test or the second derivative test.

Notice that the differentiation and algebraic manipulation in **example 2.2** is quite simple. Consequently, it does not make essential use of either the graphing or CAS technology. Yet in a CAS environment, students are very likely to use the technology, even on a simple problem.

Lesson 3

Students will use the CAS power in a variety of ways, and different calculators may create inequities.

Some students will overuse technology, and some students will avoid it. Their differing facilities at, and inclinations toward, using the technology will constitute another variable in assessment; this situation is compounded if the students do not all have access to the same technology.

Testing with Technology: Lessons Learned

The authors' experiences in their own classrooms indicate that many students both overuse and misuse technology. In 1983, we discovered that students, usually the most adept students, were overusing the scientific calculators and requiring additional time to complete the AP calculus examinations. During calculus examinations, a calculator may be needed in only a half-dozen places, but many students use the calculator constantly to perform even the most trivial arithmetic, such as dividing 10 by 5. They will do the same with CAS. If students have CAS available, they will probably not learn the expansion of $(a+b)^2$, or they will not risk their memory of the expansion. In the initial stages of CAS use, testing requires more time than previously needed, because punching the buttons is time-consuming. Students also tend to take for granted the results of the solver feature of the calculator; they report only one solution to an equation when it has two or more solutions and do not recognize the importance of the initial value of the variable for which they are solving.

Some students, especially those with less experience with technology in previous mathematics courses, may not recognize when technology is essential or helpful. They may resort to traditional methods even though these traditional methods are difficult or impossible—for example, evaluating definite integrals by antidifferentiation. Learning when and when not to use technology is a new aspect of education.

Different functionalities of calculators—CAS, built-in solvers, numerical integrators, and interoperability of functionalities—create inequities among students. We use the term *interoperability* to mean the ability to use two or more functionalities simultaneously in one command, such as using the solve feature and feeding the result into the integrating feature with a single command.

The following are two examples of examination items that "level the playing field" for students using different technologies and that focus on the concept or skill being tested. Both **example 3.1** and **example 4.1** have been reprinted with permission of the College Entrance Examination Board.

Example 3.1

On the 1995 AP calculus examinations, students were asked to analyze the behavior of the function f defined by

$$f(x) = \frac{2x}{\sqrt{x^2 + x + 1}}$$

by finding the absolute minimum value for f and justifying their answers. The derivative of f was given to the students as

$$f'(x) = \frac{x+2}{(x^2+x+1)^{3/2}}.$$

This format of this item allowed students to use the correct derivative to analyze the behavior of f, thereby eliminating the difficulties that a differentiation or algebraic mistake could cause a student in giving the kind of analysis of the behavior of f that the examiners desired. The item shown was intended to test graphical analysis and optimization using the derivative; it was not intended to test finding and simplifying derivatives, which would be tested elsewhere on the examination. Stating the derivative in the problem means that students with CAS had no advantage over students without CAS.

Example 3.2.2

On past AP calculus examinations, students have been asked to show that a derivative taken implicitly is equal to a given expression. For example, students taking the 2000 AP calculus examination were given the equation $xy^2 - x^3y = 6$ and asked to show that

$$\frac{dy}{dx} = \frac{3x^2y - y^2}{2xy - x^3}.$$

Students were next asked to write equations of tangent lines and to find points where the tangent lines were vertical.

Asking the question in this way accomplishes two goals: (1) It eliminates the advantage of having CAS, and (2) it smoothes the way for the second part of the question, which asks students to use the derivative to describe tangent lines. The tangent-line responses are likely to be easier to evaluate consistently if students use the correct symbolic derivative.

Lesson 4

Students' responses to questions will be more variable and hence less predictable.

Technology has introduced more variety in the way solutions to problems can be approached. Traditional problems in algebra and calculus had canonical solution methods that were reinforced through repeated publication in textbooks and presentations in classrooms. Teachers could predict with reasonable certainty what students would attempt to write as solutions. As teachers gain more experience with technology, which in turn becomes more uniform in its functionalities, they may someday develop new canonical solution methods but they should expect variety, at least for a while.

Lesson 5

Without the sequential nature of algebraic manipulations, students' papers will be less organized.

Mathematical reasoning, the way information is processed and represented, changes whenever it is supported by computer. See Devlin (1997) for an analysis of computer-aided logical reasoning. Mathematical reasoning can be done with pencil and paper, from graphic representations by the computer, and with CAS or numerical processes. Students have no canonical ways to record or document this combination of reasoning processes. Without the paper trail of computations, teachers may have difficulty determining and assessing the method a student used. On the one hand, students who are asked to record their reasoning processes may give no explanation for the process used, may simply present "the answer," or may explain their thinking by using such a phrase as "used my calculator." On the other hand, students may spend a great deal of time listing each keystroke used.

Students should learn to present clear and well-organized solutions to problems. Complex algebraic expansions and simplifications and such processes as intricate methods of integration have taught generations of mathematics students how to organize, how to be careful, and how to write clearly. These skills are necessary to avoid errors. Without this discipline in the mathematics curriculum, some other medium is needed to teach organization and communication. Clearly, students need to learn to write in ways that they do not presently learn.

Lesson 6

Students' responses will be more difficult to evaluate.

The nature of students' work changes with CAS. In addition to being less predictable and lacking the organizational requirements of algebraic manipulations, students' responses are personalized by their language and experiences. Students' interpretations and justifications of mathematical situations are often imprecise and disorganized. Consequently, more effort is required, on the part of the evaluator, to read and understand what students mean and to evaluate responses fully and fairly. **Example 6.1**, reprinted with permission of the College Entrance Examination Board, illustrates the decisions that evaluators must make regarding what students mean to say with language or diagrams.

Example 6.1

A question on the 1998 AP calculus examination asked the student to find the absolute minimum value of the function f defined by $f(x) = 2xe^{2x}$ and to justify the answer.

The student finds that the derivative $f'(x) = 2e^{2x} + 4xe^{2x}$ has only one zero, occurring at $x = -\frac{1}{2}$. So the absolute minimum is

$$f(-\frac{1}{2}) = -\frac{1}{e}.$$

Students' justifications were difficult to read and evaluate because many were cryptic, disorganized, and imprecise. One that appeared frequently was the following diagram:

$$f'(x) \text{ -------} -\frac{1}{2} \text{ + + + + + + + +}$$

Does the student mean to say that the derivative is negative for all real numbers less than $-½$ and positive for all real numbers greater than $-½$, or is the student just showing the sign of the derivative for values of x near $-½$? Students commonly make the error of noting the existence of a local minimum at $-½$ and give it as justification of an absolute, or global, minimum. The following question arises: Do the arrows on the ends of the line segment serve as the quantifier "for all"? In this case, the readers decided that the answer was "yes." Undoubtedly, some students meant the arrows to serve as quantifiers, whereas some did not. With CAS, many more points on examinations will be attached to such justifications as the one shown in **example 6.1**.

Lesson 7

Examination questions will be more conceptual and, hence, more difficult for some students.

Because of the shift in values from algebraic and differentiation skills to explanations of concepts, justifications, and interpretations, different students will have different degrees of difficulty in examinations using CAS. Although some students find the conceptually oriented courses easier, many students will find them more difficult (Monaghan 2000). Algebraic processes have been the "lifeline" of some students. Therefore, their first experience of not having those algebraic processes available for credit is likely to result in lower scores.

Lesson 8

Students will become less proficient at algebra and arithmetic manipulations and will memorize fewer formulas and other facts.

As we observed in the discussion of lesson 3, many students will use CAS for arithmetic operations, algebraic simplifications, and expansions. Even if students are tested without calculators part of the time, they will get less experience using algebra and arithmetic.

If students have access to CAS, then they have a virtual library of facts about the course; these facts include trigonometric identities, differentiation formulas, and measurement formulas. Consequently, students will probably read basic facts from the machine. Specific facts that need to be remembered should be tested without students' access to calculators. Thus, tests designed for calculator use can operate, and rightly so, under a "no holds barred" policy regarding what can be stored in the calculator. Recognizing the need for both types of tests, the AP calculus examinations have parts that require a calculator, with whatever data the student has available in the calculator; the examinations also have parts that prohibit the use of any calculator.

Everyone seems to agree on the existence of a basic body of facts and skills that every student should know without using a calculator, but educators cannot agree on what comprises this body of facts and skills.

Lesson 9

Some areas of mathematics are affected more than others.

CAS technology effects algebra, differentiation, and antidifferentiation items significantly more then it affects graphical analysis, geometry, numerical methods, and modeling items. In such areas as algebra, differentiation, and antidifferentiation, students with CAS have a significant advantage over students with graphic-numeric technology. Students can produce the Taylor coefficients for power series by using CAS but not by using graphic-numeric technology; they can also use CAS to solve many differential equations. Even in the process of justifying answers, students can use CAS to help. For example, if a student can evaluate an improper integral or an infinite series with a CAS, then that student knows whether to try to justify *convergence* or *divergence*.

Lesson 10

Difficult questions about forms of answers will arise.

The most difficult question posed by the use of CAS is "What do you mean by an answer?" Actually, this question has always been difficult to answer, but we previously avoided it because the mathematics community seemed to understand that some answers were "better" than others. For example, $x^2 + 5$ is a better solution to the problem "Solve the differential equation $y' = 2x$ with $y(0) = 5$" than is

$$y = 5 + \int_0^x 2t\,dt.$$

Additional examples of such difficulties follow.

On the 1997 AP calculus examination, the scoring of one question, reprinted here with permission of the College Entrance Examination Board, did not depend on simplification of

$$5\sin\frac{5\pi}{2}$$

to 5 but did require that the elliptic integral

$$\int_{1.25}^{1.75} \sqrt{9\pi^2 \sin^2(\pi t) + 25\pi^2 \cos^2(\pi t)}\,dt$$

be simplified to 5.392, thus raising questions that are not easily answered. First, both

$$5\sin\frac{5\pi}{2}$$

and

$$\int_{1.25}^{1.75} \sqrt{9\pi^2 \sin^2(\pi t) + 25\pi^2 \cos^2(\pi t)}\,dt$$

are numbers, and both can be approximated by a decimal number using a calculator, the elliptic integral being more difficult only because it requires more keystrokes to input. Then why should one be left as is and the other require a decimal approximation? Why is the integral approximated by 5.392 rather than by 5.4 or 5.39213?

In a more fundamental, simpler example, should students write their final answer as

$$\sin\frac{\pi}{3}$$

or

$$\frac{\sqrt{3}}{2}$$

or 0.866? Does it matter? Should they be required to know the exact form of

$$\sin\frac{\pi}{3}$$

as opposed to

$$\sin\frac{\pi}{5},$$

which is

$$\frac{\sqrt{10-\sqrt{20}}}{4}?$$

What accuracy in decimal answers should be required? Encouraged?

The instructions to the free-response section of the AP calculus examinations include this statement: *Whenever you use decimal approximations, your answer should be correct to three decimal places after the decimal point.* In practice, the statement "correct to three decimal places" has been interpreted at the Reading to mean *correctly rounded to three places or truncated to three places.* No compelling, logical reason can be cited to explain why accuracy to three decimal places was chosen; direction had to be given, so this direction was chosen. When challenged by critics, the explanation was "We know that π and 3.142 are different numbers, but when we ask students to give the decimal name of π, we require that they give us only the name up to three digits after the decimal point."

Several issues surround this issue of accuracy. An overriding issue on a national examination is that the directions must be clear. For example, relative errors are not considered, but rather, the absolute error. The concept of relative error is probably not well understood in the student population. One item on the 1996 AP calculus examinations, reprinted as **example 10.1** with permission of the College Entrance Examination Board, illustrates this issue and other related issues.

Example 10.1

The rate of consumption of cola in the United States is given by $S(t) = Ce^{kt}$, where S is measured in billions of gallons per year and t is measured in years from the beginning of 1980.

(a) The consumption rate doubles every 5 years and the consumption rate at the beginning of 1980 was 6 billion gallons per year. Find C and k.

(b) Find the average rate of consumption of cola over the 10-year time period beginning January 1, 1983. Indicate units of measure.

The choice of units, billions of gallons per year, proved to be troublesome. In part (a), the student was expected to find $C = 6$ and

$$k = \frac{\ln 2}{5},$$

the latter being acceptable given either as 0.138 (truncated) or 0.139 (rounded). Some students changed units to gallons per year, changing the value of C to 6,000,000,000, hence changing the definition of the given function S. This response sparked much discussion among participants in the Reading. If students changed units to gallons per year, they had no chance to compute the average rate in part (b) accurate to three decimal places because such a calculation would require carrying fifteen places in the computation.

If a student used $k = 0.138$, which is considered a correct answer to part (a) of **example 10.1**, then the answer to part (b) correct to the required three places after the decimal, is 19.567 billion gallons per year. The more accurate answer is 19.680 billion gallons per year. Thus the student's answer is not correct in the first place after the decimal, let alone the third place, even when staying with units in billions of gallons. (Intermediate rounding is a major issue; a more in-depth discussion of the problem of intermediate use of decimals can be found in the next example.) If the student used $k = 0.139$, a similar situation occurs. An error in the third decimal place after the decimal sounds rather small, but when the units are billions of gallons, the third decimal place represents millions of gallons. The relative error is not affected by the change in units.

Below is another example, used with permission of the College Entrance Examination Board, in which the desired accuracy of the answer is not obtained if the student uses the same degree of accuracy in the intermediate steps.

Example 10.2

A question on the 1997 AP calculus examinations had a part that required the student to find x in

$$\left[0, \frac{\pi}{2}\right]$$

satisfying

$$\sin x = \frac{2}{\pi} = 0.6366\ldots$$

The answer, correct to three decimal places, rounding or truncating, is 0.690. Is the final 0 required, or is 0.69 satisfactory? However, the student using

$$\frac{2}{\pi} = 0.637$$

obtains $x = 0.691$ (rounding). Furthermore, recall the discussion in **example 2.1**. We saw that intermediate rounding to one decimal place still produced an answer that was correct. Which solution represents better work, the solution with intermediate rounding to one decimal place or the solution with intermediate rounding to three decimal places? The former can lead to correct accuracy in the final answer, whereas the latter may not.

Unfortunately, understanding the errors introduced by intermediate rounding requires more analysis than can be covered in a calculus course. What, then, should students to be told to do? Never round? Carry all the precision allowed by the calculator? When asked for an intermediate answer to a certain accuracy, should students be allowed to use—without penalty—the intermediate number to the requested accuracy in computation, even if the rounding introduces too much error in a subsequent answer? As the preceding examples illustrate, the requested accuracy of three decimal places is not always enough to yield that accuracy in subsequent computations.

Another issue arises, a result of the internal accuracy of the calculators themselves. How should we deal with lack of the requested accuracy in answers produced by built-in "solve" keys on calculators or by any other "black box" that the student trusts to provide correct answers?

Example 10.3

In an item on the 1997 AP calculus examination, reported here with permission of the College Entrance Examination Board, the students were asked to evaluate the integral of $\sqrt{x-3}$ from 3 to 6.

One calculator with a built-in definite-integral key gave the answer 3.463 rather than the acceptably correct 3.464. Note the steepness of the graph near $x = 3$; this danger was alluded to earlier. Should the answer 3.463 receive full credit because the student believes the "black box" calculator?

If the answer is yes, then what should we do about the student who is told by a teacher to always use the trapezoidal rule—programmed in—with twenty equal subdivisions? When grading a student's paper, how does the reader know the source of the bad approximation? Does the error made by the calculator company deserve more consideration than one made by a calculus teacher? Both are "black boxes" to the student. The Chief Reader decided at the Reading to give full credit for the inaccurate calculator answer and was sympathetic to extending the same tolerance to the student who was misled by a teacher. In other contexts, the AP program may be misleading the students by considering accuracy to three decimal places as correct. Unfortunately, information given to students by their teachers is not cast in plastic and ceramics like the calculator, so readers have difficulty surmising the source by looking at students' work.

We extend the "black box" discussion a little further. Should students be given credit for answers obtained from a calculator in the incorrect mode? Most of us would answer *no*, particularly if students obtained incorrect answers by calculating values of the trigonometric functions when the calculator was in degree mode. Should students enrolled in a course studying the calculus of real variables be given credit for finding the value of an integral with the calculator in complex mode?

Example 10.4

The integral

$$\int_{-1}^{8} \sqrt[3]{x}\, dx$$

is evaluated as

$$\frac{45}{4}$$

by a certain calculator when it is in "real" mode, but in "rectangular" mode the evaluation becomes

$$\frac{99}{8} + \frac{3\sqrt{3}}{8} i.$$

The sophisticated CAS Mathematica reports the answer to example 10.4 to be

$$12 + \frac{3}{4}(-1)^{1/3},$$

leaving it to the user to decide the value of

$$(-1)^{1/3}.$$

This latter form is even more intriguing than the earlier discussion of

$$\sin\frac{\pi}{3};$$

indeed, the refusal of some calculators to graph the cube-root function has affected the way some AP calculus questions have been written.

Mathematics Faculty Attitudes toward CAS

Although questions of equity and implementation existed when the representatives of the AP program began contemplating the calculator issue, most of the high school teachers associated with the AP program seem to have adapted to present policy and many are enthusiastic users of calculators in the classroom. The picture in collegiate classrooms is not as uniform. Most of the current collegiate mathematics faculty did not learn mathematics in a CAS environment and have no personal experience with the effects of doing so. Most collegiate mathematics faculty members were very strong mathematics students and very able at symbolic manipulation. Their talents and skills have served them well, and they have known students whose limited symbolic manipulation skills have been the students' undoing. Consequently, they put considerable stock in the value of algebraic skills. To many of them, these skills are basic, like language.

Yet the ready availability of CAS for the last twelve years has affected some mathematical research because many mathematicians use CAS as an aid in performing difficult and complex algebraic simplifications or computations needed for their own advanced work. Why do some college faculty members believe that CAS is an excellent tool for them and not for their students? Very simply, they believe that students must "walk before they run," as one colleague said. By running, he meant using the speedy CAS. By walking, he meant doing algebra by hand. But the extent to which by-hand skills are deemed necessary does change with time. College algebra textbooks of the early twentieth century are very different from those published in the latter part of the century. Mathematicians of the authors' generation grew up using textbooks that had logarithm and trigonometric tables in the back of the book; they also performed drills involving linear interpolation. The widespread availability of calculators has rendered such drills obsolete, and such tables no longer appear in all textbooks. How much time will calculus students of the future spend on by-hand techniques of antidifferentiation? Will the ready availability of CAS mean that our calculus textbooks will no longer include a table of integrals in the future?

Disclaimer

The opinions expressed in this chapter do not necessarily represent positions of either The College Board or Educational Testing Service.

REFERENCES

Crossroads in Mathematics: Standards for Introductory College Mathematics Before Calculus. Memphis, TN: American Mathematical Association of Two-Year Colleges, 1995.

Devlin, Keith. "The Logical Structure of Computer-Aided Mathematical Reasoning." *American Mathematical Monthly* 104 (August–September 1997): 632–46.

Hurley, James F., Uwe Koehn, and Susan L. Ganter. "Effects of Calculus Reform: Local and National." *American Mathematical Monthly* 106 (November 1999): 800–811.

Madison, Bernard L. "Calculators and AP Calculus: Observations and Questions." In *Proceedings of the 1997 Technology Transitions Calculus Conference*, Dallas, Texas, forthcoming.

Monaghan, John. "Some Issues Surrounding the Use of Algebraic Calculators in Traditional Examinations." *International Journal of Mathematics Education in Science and Technology* 31, no. 3 (May 2000): 381–92.

Morgan, Rick, and Joe Stevens. "Experimental Study of the Effects of Calculator Use on the Advanced Placement Calculus Examinations." Educational Testing Service Research Report RR-91-5. Princeton, NJ: Educational Testing Service 1991.

National Council of Teachers of Mathematics (NCTM). *Curriculum and Evaluation Standards for School Mathematics*. Reston, Va.: NCTM, 1989.

Steen, Lynn Arthur, ed. *Calculus for a New Century: A Pump, Not a Filter*. MAA Notes No. 8. Washington, D.C.: Mathematical Association of America, 1988.

CHAPTER 18

Traditional Assessment and Computer Algebra Systems

Lin McMullin

COMPUTER algebra systems (CAS) offer many opportunities for deeper and more realistic "doing" of mathematics. Using CAS for explorations, projects, reports, and experiments, as well as for working with numbers, functions, and situations that are simply too complicated to do by hand, pushes the mathematics curriculum forward. The fact that mathematics in the "real world" is done with CAS pulls it forward. If we accept, as this writer does, that CAS belong in mathematics classrooms, at least from the beginning of the study of algebra, how do we use them to encourage learning and how do we assess that learning?

Various alternative methods of assessment come to mind, but traditional assessment—questions with short answers, multiple-choice questions, and questions requiring a few lines of work—will not disappear. Even if we could avoid traditional assessment methods in our classrooms, we could not ignore them forever because district, state, and national examinations will be with us for a long time. Assessment drives curriculum. The skills and concepts that will be on the test must be taught. If district, state, and national tests require the use of technology and computer algebra systems, then teachers will teach their students to use them. Arguments surrounding the use of CAS may continue, but CAS use will be taught.

Assume that your students have CAS available and that you, your district, and your state will test students' abilities to do mathematics with CAS technology. Assume also that you want your students to know how to do some things *without* technology. How do you balance the two goals within a traditional assessment format?

329

Not a Totally New Challenge

When graphing calculators first became available, The College Board was faced with decisions surrounding the use of this technology, on the Advanced Placement (AP) Calculus examinations—the first national group to encounter such a dilemma. Although graphing calculators do not usually have CAS capabilities, the challenges that CAS place on testing are similar to the challenges experienced by The College Board and Educational Testing Service (ETS) in administering and scoring the AP Calculus examinations. Their experiences with graphing calculators provide us with a starting point for thinking about traditional assessment with CAS.

Members of the committee that writes the AP Calculus examinations are very familiar with the capabilities of *all* the graphing calculators available. This knowledge allows them to write questions that do not give students having more sophisticated calculators an advantage over other students. Teachers at the school level, as well as the state and national levels, must be familiar with the CAS their students are using. As a rule of thumb, teachers should assume that CAS can do *all* the symbolic manipulation, as well as graphical and numerical work, that a secondary school student will need do in high school—and then some. Teachers cannot assume that because they have not taught a particular CAS procedure, the students have not figured it out by themselves.

AP Calculus students are expected to know how to do four things with their graphing calculators. Although graphing calculators can do much more, students are expected to be able to use them to—

1. solve an equation numerically,

2. graph a function in a given window,

3. find a numerical approximation for a derivative at a point, and

4. find a numerical approximation for a definite integral.

Teachers are expected to teach the four calculator skills listed above, and students are expected to learn these skills. These four skills are specifically tested on the AP Calculus examination; that is, questions related to these skills appear that are very difficult or impossible to do without technology. Furthermore, students are expected to use these skills when appropriate; they are not given extra credit for doing the problems by hand. For example, once a definite integral is set up, students are expected to evaluate it with a calculator; no additional credit is awarded for writing the antiderivative, and, in fact, credit may be deducted if students make a mistake in writing the antiderivative.

Students are not permitted to use their calculators on the examinations for any of the other tasks that the calculators can accomplish. This restriction is accomplished in three ways. First, the examinations include sections for which calculators are not permitted. Those skills, concepts, and procedures that are judged to be important enough for the student to know how to do without a calculator are tested on the noncalculator portions of the test. Second, the examinations include questions for which graphing calculators are of little or no apparent use. A typical question of this type would require an analysis of a function, given the graph of its derivative with no symbolic rule to manipulate. Third, the examinations include free-response questions that require students to show their work. A typical problem may require a definite integral for its solution. The integral is not given, so the student must write it in standard notation to receive full credit. Only the evaluation of the integral may be done with the technology. Topics for which the evaluation of definite integrals by hand is judged to be an important skill are tested separately on the noncalculator portion of the examination. The considerations below—taken from experiences of the AP Calculus examination system—are useful in pointing out ways to use CAS in classroom assessment across the mathematics curriculum.

1. The teachers and evaluators who write the tests in a school, a state, or the nation must be familiar with the capabilities of the CAS that students use.

2. Those technological skills that students are allowed to perform with the CAS should be specifically delineated and carefully specified for teachers and students. The teachers must teach the skills; the students must learn them. The technological skills should be tested. The technology that is permitted and prohibited may change during a school year and across the grades. A teacher may want to assess students' ability to perform certain skills without a CAS before allowing them to use a CAS. Certain topics may be introduced through CAS-based activities, with students doing the corresponding paper-and-pencil work later.

3. Skills that students are allowed to do with a CAS should be tested using questions specifically designed for this purpose.

4. Skills that students are not permitted to do with a CAS should be tested using questions for which a CAS cannot help them or by using tests or sections of tests on which a CAS is not permitted.

Students learn to do manipulations by doing them. In the process of performing manipulations, students may perceive that the manipulation itself is the important thing, but it is not. *When* to do the manipulation, *why* to do it, *what* the before and after forms mean, and *what* they are used for are far more important than the manipulation skills. Instructional strategies that reflect the values and assessment guidelines given previously can be illustrated well by two examples from beginning algebra: (1) expanding and factoring polynomials (2) combining like terms.

Expanding and Factoring Polynomials

Expanding and factoring of polynomials are central topics in the algebra curriculum. Expanded and factored forms—each used in certain situations—provide different information about a polynomial. Knowing when to expand, when to factor, and how to use each form are important parts of algebraic understanding. The actual skills of expanding and factoring by hand are much less important. Students should spend a few days expanding polynomials and a few days factoring simple trinomials to understand the relationship between the expanded and factored forms. Perhaps students should also be able to do, in their heads, the most common two types of expanding and factoring—the difference of perfect squares and perfect square trinomials. Having learned these concepts, students can be tested without access to CAS.

Solving equations by factoring is the most common topic to follow simple factoring, and it is a good place to introduce the CAS factoring utility. Students may be allowed to do the factoring with the CAS and required to show all the other work. The advantage of using the CAS is that polynomials of degree higher than two can be investigated, making unnecessary the postponement of polynomials of higher degree until students take a precalculus course. First-year algebra students can now be expected to use factoring to solve the equation

$$x^5 + x^4 - 18x^3 + 14x^2 + 17x - 15 = 0.$$

Although the CAS could solve such equations in one line, students could be required to show their work, doing only the factoring with their CAS:

$$x^5 + x^4 - 18x^3 + 14x^2 + 17x - 15 = 0$$

$$(x - 3)(x - 1)^2 (x + 1)(x + 5) = 0$$

$$x - 3 = 0 \text{ or } x - 1 = 0 \text{ or } x + 1 = 0 \text{ or } x + 5 = 0$$

$$x = 3, x = 1, x = -1, x = -5$$

When the solution is required in a more advanced situation at a later date, it may be found directly without showing the factoring. Rational expressions also require factoring for simplification. The CAS usually simplifies rational expressions automatically, but by being required to show the other work, students could be limited to doing only the factoring with a

CAS; for example,

$$\frac{x^2+4x-5}{x-5} \div \frac{x^2-4x+3}{x^2-8x+15} = \frac{x^2+4x-5}{x-5} \cdot \frac{x^2-8x+15}{x^2-4x+3}$$

$$= \frac{(x-1)(x+5)}{(x-5)} \cdot \frac{(x-5)(x-3)}{(x-3)(x-1)}$$

$$= (x+5).$$

Later in the year, or in following years, students could be allowed to present only the simplified form in the context of a longer problem.

Adding Like Terms

As a second example from beginning algebra, consider the skill of adding like terms. Teachers may ask students to enter on their CAS a few problems, such as $7 + 7 + 7 + 7 + 7$ and $x + x + x + x + x$, and look at the results returned, 35 and $5x$, respectively. They may allow this exploration to promote to a discussion of the concept of multiplication as repeated addition, emphasizing that adding five of the same number or variable is the same as multiplying 5 times the number or variable. Next, they may progress to such examples as $a + a + a + b + b = 3a + 2b$, $5x + 3y + 7x - 2y = 12x + y$, and $6b - b = 5b$, each time introducing an example and carefully discussing why the CAS is returning the answer displayed on the screen. Teachers may decide to first assess students without CAS. When students encounter such expressions later on, the CAS can be used to simplify automatically, but not mysteriously.

As the Year Goes On

What is prohibited on one test may be permitted on subsequent tests. Care must be taken in this kind of progression. Manipulation skills need be taught only to the extent that they help the student see and understand what is happening. Requiring students to become extremely facile at some technique, only to say, "You will never have to do that again, because from now on your CAS can do it for you" would waste time and have a deleterious effect on motivation.

The list of skills that a student is permitted to do using a CAS will grow during the year. I suggest that teachers require students to work by hand for a short period of time on most, if not all, the skills they learn in high school mathematics, so that they develop an understanding of what is going on rather than become experts. After working by hand, teachers should allow students to use CAS in contexts in which the technology is useful.

Curriculum Changes

Just as scientific calculators led to the removal of computations with logarithms from the curriculum, CAS will cause some topics, such as the quadratic formula, to disappear from the curriculum. For example, CAS can give exact real, surd, and complex solutions to quadratic equations. On the one hand, students may need to understand where surd and complex roots come from, and completing the square may be instructive in achieving this goal. On the other hand, students will probably not have any use for the quadratic formula once they have a CAS.

Interspersed between the chapters of this book are activities that illustrate how CAS can be used to present the solution of secondary level problems. The copies of the CAS screen, as well as the written explanations under them, comprise the solution. Notice that in each activity, all the "work," except the actual symbol manipulation, is shown, so that what the student is doing and thinking is clear. Using only problems with "nice" numbers is unnecessary—any numbers or variables work well. Without showing the symbol manipulation, the solutions look much different from what is currently expected.

Conclusion

Many things need to change to accommodate CAS. This necessity is a good thing. In mathematics, CAS can do many things better than humans, but we cannot simply hand students CAS and let them go. Only when CAS use by students is required on traditional assessments at the local, state, and national level will CAS truly become part of the curriculum—a change that must occur. Done carefully and purposefully, traditional assessment can lead the way to the inclusion of the use of this important new technology in the high school mathematics curriculum and increase the amount of real mathematics that students can do.

Activity 11

Prove that the graph of every cubic polynomial is symmetric about its point of inflection.

Solution:

- $a \cdot x^3 + b \cdot x^2 + c \cdot x + d \to f(x)$ Done

- $\text{solve}\left(\dfrac{d^2}{dx^2}(f(x)) = 0, x\right)$ $x = \dfrac{^-b}{3 \cdot a}$

- $f\left(x + \dfrac{^-b}{3 \cdot a}\right) - f\left(\dfrac{^-b}{3 \cdot a}\right) \to f1(x)$ Done

- $f1(x) = {}^-f1({}^-x)$ true

The *first line* defines the general cubic and names it $f(x)$.

The *second line* finds the x-coordinate of the point of inflection: the zero of the second derivative of $f(x)$.

The *third line* translates $f(x)$ so that the point of inflection,

$$\left(-\dfrac{b}{3a}, f\left(-\dfrac{b}{3a}\right)\right),$$

is at the origin and calls this new function $f1(x)$.

The *fourth line* shows that the equation $f1(x) = -f1(-x)$ is an identity (always true).

We conclude that the translated function is an odd function and symmetric to the origin. Therefore the original $f(x)$ is symmetric to its point of inflection. QED.

Note: Two questions:

   (a) The coordinates of the point of inflection are

$$\left(-\frac{b}{3a}, -\frac{bc}{3a} + \frac{2b^3}{27a^2} + d\right)$$

and

$$f1(x) = \frac{x\left(3a^2x^2 + 3ac - b^2\right)}{3a}.$$

We shudder at the thought of anyone's doing these calculations by hand. We did not; they are available on the CAS if we need them. Likewise, we could see the first and second derivatives, but do we really need to see them?

(b) Students who can produce the screen shown above know what they are doing. Do we really need to make these students perform the procedure by hand to be convinced that they understand the mathematical concept?

The alternative solution below is more complicated but perhaps more elegant than the solution above.

- $a \cdot x^3 + b \cdot x^2 + c \cdot x + d \rightarrow f(x)$ Done

- $\text{solve}\left(\frac{d^2}{dx^2}(f(x)) = 0, x\right)$ $x = \frac{^-b}{3 \cdot a}$

- $f\left(\frac{^-b}{3 \cdot a}\right) - f\left(\frac{^-b}{3 \cdot a} - k\right) = f\left(\frac{^-b}{3 \cdot a} + k\right) - f\left(\frac{^-b}{3 \cdot a}\right)$ true

The third line compares the y-coordinates of points k units to the right and left of the inflection point. It shows that these points are always the same distance below and above the inflection point, thus establishing the symmetry.